d.PLUS 立体城市研究院书系

2012"亚洲垂直城市"国际设计竞赛暨研讨会
Vertical Cities Asia International Design
Competition & Symposium

每个人都会变老
EVERYONE AGES

U+E

广西师范大学出版社
桂林

图书在版编目(CIP)数据

每个人都会变老 / U+E主编. – 桂林：广西师范大学出版社，2013.7
ISBN 978-7-5495-3788-4
（d-PLUS 立体城市研究院书系）

Ⅰ.①每… Ⅱ.①U… Ⅲ.①城市建筑 – 建筑设计 – 研究 Ⅳ.①TU984

中国版本图书馆CIP数据核字(2013)第105620号

出品人：刘广汉
责任编辑：周丹

广西师范大学出版社出版发行
广西桂林市中华路22号　邮政编码：541001
　网址：http://www.bbtpress.com
出版人：何林夏
全国新华书店经销
销售热线：021-31260822-882/883
北京建宏印刷有限公司
（地址：北京市顺义区后沙峪吉祥工业区吉安路2号　邮政编码：100318）

开本：787mm×1092mm　　1/8
印张：18.5　　　　　　　字数：13千字
2013年7月第1版　　　　2013年7月第1次印刷
定价：296.00元

――――――――――――――――――――――
版权所有　翻印必究
如发现印装质量问题，影响阅读，可寄本社退换。
（电话：010 – 68888588）

荣誉出品人 Honorable Producer
冯仑 Feng Lun
郝杰斌 Hao Jiebin

出品顾问 Publishing Consultant
李娜 Li Na
刘刚 Liu Gang

出品人 Producer
葉春曦 Isa Ye

主编 Chief Editor
于冰 Yu Bing

项目总监 Project Manager
英小盟 Christy Ying

项目策划 Project Planner
林贺佳 Lin Hejia
汪洋 Wang Yang

编辑总监 Managing Editor
吴博 Wu Bo

客座编辑 Guest Editor
王才强（新加坡）
Heng Chye Kiang (Singapore)
伍伟坚（新加坡）
Ng Wai Keen (Singapore)

编辑 Editor
张雨辰 Zhang Yuchen
杨毅 Yang Yi

美术编辑 Art Editor
逄苹 Pang Ping
李渔 Li Yu
黄洁 Huang Jie

特别鸣谢 Acknowledgement

评审团成员 Jury Members

李灿辉 Tunney F. LEE

出生于中国广东，在波士顿唐人街长大。他获得了密歇根大学的建筑学学士学位，曾在罗马大学做富布莱特访问学者。他曾担任麻省理工学院的城市研究与规划系的系主任（现在是建筑和城市规划系的名誉教授），且参与创建了香港中文大学建筑系。

Tunney F. Lee was born in Guangdong, China and grew up in Boston's Chinatown. He has a Bachelor of Architecture from the University of Michigan and was a Fulbright Fellow at the University of Rome. He was Head of Massachusetts Institute of Technology(MIT)'s Department of Urban Studies and Planning (now Professor Emeritus of Architecture and City Planning) and the founder of the Department of Architecture at the Chinese University of Hong Kong.

David MANGIN

建筑师和城市规划师。2004年他提出的巴黎Les Halles区域改造的竞争方案被巴黎市长Bertrand Delanoë选中。2008年，他还赢得了法国城建大奖。Mangin在巴黎开有一家工作室（Seura Architectes Urbanistes）并且参与了"大巴黎"计划，此外还任教于Marne-La-Vallée建筑学院和ENPC学院。

David Mangin is an architect and town planner. His competition schemefor redeveloping the Les Halles area of Paris was selected by the city's mayor, Bertrand Delanoë, in 2004. He was also the winner of the Grand Prix de l'Urbanisme Français in 2008. Mangin runs an office (Seura Architectes Urbanistes) in Paris and is involved in the Grand Paris Programmes.He teaches in the School of Architecture of Marne-La-Vallée as well as at ENPC School (High School of Engineers).

Joaquim SABATÉ BEL

Joaquim Sabaté Bel在西班牙加泰罗尼亚职业学院（UPC）的城市规划学院担任院长和教授，也是UPC博士、城市化研究硕士和区域设计方案的协调员。Sabaté Bel在多个欧洲和南美国家负责城市设计、城市和区域规划。他曾三次获得西班牙城市化国家奖；此外，还获得了加泰罗尼亚城市化奖和遗产项目特别奖。

Joaquim Sabaté Bel is Professor and Chair of Town Planning at the Polytechnic University of Catalonia(UPC) in Spain. He is the coordinator of the PhD, Research Master in Urbanism, and Regional Design programmes at UPC.
Sabaté Belis responsible for urban designs, as well as urban and regional plans, in different European andSouth American countries. He has been awarded the Spanish National Prize of Urbanism three times; the Catalonian Prize of Urbanism; and the Special Prize for Heritage Projects.

陈青松 TAN Cheng Siong

建筑师和Archurban集团（新加坡和深圳）的创始人。他获得过建筑学学位（1964）和城市规划硕士学位（1972）。从1986年以来他一直担任深圳发展委员会的顾问。从1990年代中期开始陈青松向中国引入了公寓住房的概念，并且一直在深圳、杭州、武汉、上海、三亚、北京以及其他城市设计许多领先的地标性建筑。

Tan Cheng Siong is an architect and the founder of the Archurban Group (Singapore and Shenzhen). He holds a Diploma in Architecture (1964) and a M.A. Urban Planning (1972).Serving as an advisor to the Planning Committee of Shenzhen since 1986, Tan introduced the concept of condominium housing to China in the mid-1990s and continues to design pioneering landmarks in Shenzhen, Hangzhou, Wuhan, Shanghai, Sanya, Beijing, and other cities there.

严迅奇 Rocco YIM

出生于香港并在香港接受教育，目前任香港许李严建筑师事务有限公司执行董事。严迅奇先生常应邀到当地和全球各地的各种研讨会和座谈会发表演讲。目前他是香港大学（HKU）建筑系的咨询委员会成员，香港大学专业进修学院（HKUSPACE）的客座教授，香港康乐及文化事务署博物馆项目顾问。

Rocco Yim was born and educated in Hong Kong, and is currently Executive Director of Rocco Design Architects Limited in the same city. Yimis regularly invited to speak atlocal and international symposia and seminars. He is currently on the Advisory Council for the University of Hong Kong (HKU) Department of Architecture, is Adjunct Professor at HKU SPACE, and is a Museum Adviser to the Leisure and Cultural Services Department (LCSD).

目录 Contents

06　**序言 Preface**
　　寻找未来幸福的阶梯
　　Towards a Sustainable and Happy Urban Future

08　**新加坡国立大学设计与环境学院院长王才强教授寄语**
　　The Foreword from Prof. Heng Chye Kiang, Dean of School of Design and Environment, National University of Singapore

10　**大赛介绍 Introduction**
　　2012年亚洲"垂直城市"国际竞赛以"每个人都会变老"为主题，设计对象是韩国首尔龙山区的一个一平方公里的区域。该竞赛旨在鼓励参赛队伍提出积极的老龄化应对方案，解决应如何抓住机遇、在维持现有城市功能水平的同时保障居民的终身福祉。设计方案考虑"积极老龄化"及"居家养老"等概念，为年老及年青的一代设计适当的照护环境。
　　This year's competition explores the theme "Everyone Ages" and the design target place is located in one square kilometre of land in Yongshan district of Seoul. The competition explores and addresses this concern by encouraging new positive approaches to ageing society that identify opportunities for maintaining capacities and well-being over the life course. By observing concepts such as "active ageing" and "ageing in place", the solutions aim to develop appropriate built environments, for both older and younger generations; a crucial element to successful ageing within the community.

17　**参赛作品 Competition Entities**
　　在2012年"亚洲垂直城市"大赛中，来自亚洲、欧洲及美国的参赛学生，提出全面的可持续都市规划模式，以舒缓亚洲城市老龄化的问题。其中，来自荷兰代尔夫特理工大学两支参赛队伍巧用创意解决城市化问题，在国际设计比赛中双双摘下桂冠，东京大学的B队和上海同济大学的A队则分别荣获第二和第三名。
　　虽然最终只有8件作品获得了奖项，但其实每一件作品都呈现出周详的设计方式，并以全面的角度考虑各种因素，同时又展现了城市建筑远景。尽管最终的设计结果和表现形式很重要，但评审委员也积极肯定了参赛队伍们在方案中所投入的心血和思考过程。
　　In the Vertical Cities Asia 2012, students from Asia, Europe and the US offer holistic solutions to address issues relating to sustainability and aging population in Asian cities. Two teams from Delft University of Technology win International Design Competition with innovative solutions to address urbanisation problems. The second-prize winning scheme is from the University of Tokyo, Team B and the third-prize goes to Tongji University, Team A.
　　Although there are only 8 works which get prizes finally, other 12 works also distinguish a thoughtful process oriented approach, a comprehensive consideration of a variety of dimensions that contribute to city building with a clear vision towards the future. Results and form really matters, but the jury wanted to highlight the attention paid to process and programmes in the proposals.

55　**研讨论文 Symposium Papers**
　　对城市背景下的老龄化问题进行的讨论倾向于鼓励使用两种调查方法。在2012年亚洲垂直城市研讨会上公开的这两种思路穿插了对垂直度和密度进行的更全面的讨论。研讨会文件分为四个宽泛的主题分类：立体发展方向；人口老龄化空间和建筑老龄化；密度研究；案例分析。这四个主题分别提供了关于亚洲和更远地区的城市发展的有价值的见解。
　　A discussion of ageing in the context of cities tends to encourage two lines of enquiry. Both threads surfaced at the Vertical Cities Asia Symposium 2012, interspersed with more general discussions of verticality and density. The symposium papers fall into four broad subject categories that offer valuable insights into urban development in Asia and further afield: aspects of vertical development; spaces for the ageing of people, and the ageing of buildings; the study of density; and project discussions.

146　**跋 Afterword**
　　北京万通立体之城投资有限公司首席执行官郝杰斌 后记
　　The afterword from Roy Yao, the CEO of Beijing Vantone Citylogic Investment Corporation

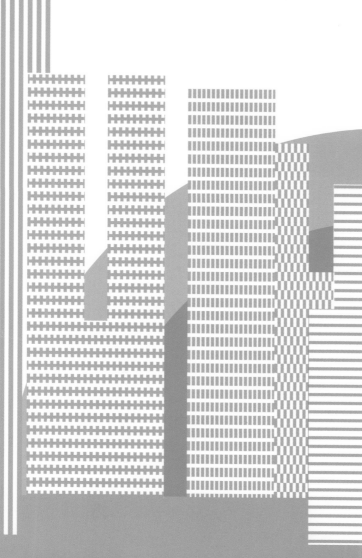

VERTICAL CITIES ASIA

竞赛主题
2011/每个人都需要新鲜空气；2012/每个人都会变老；2013/每个人都丰收满盈。

竞赛目的
竞赛每一年就特定主题针对一个快速发展同时也面临着可持续性和生活质量问题的亚洲城市展开，期待寻找到一个符合主题要求的整体方案或新型城市范例。

竞赛方式
参赛者需于一平方公里的土地为100,000人口进行规划设计，同时以亚洲城市急剧发展为前提，考虑密度、生活质量及可持续发展等要素。

设计考量
设计方案应全面考量以下方面要素：可持续发展(环境)、生活质量(包容性及社区)、可行性(可建性、财务及社会支持)、切题度(地区性、状况、气候及文化)及科技创新(科技及技术)。

团队参与
邀请全球10所知名大学参与，每所大学将选取两份设计方案参赛，且有1名教员和2名学生获得赞助出席评委会会议和研讨会。所有大学应将设计方案和文件进行分组，标明"A组"或"B组"。主办方将邀请评委会成员在研讨会上针对密集实现城市化的亚洲城市背景，以每期特定主题发表一篇论文。两名学生将代表各自团队，向国际评委会展示设计作品。

评审机制
设计评委会将通过评委会会议对所有参赛作品进行评审。评委会会议将对外开放，邀请所有教员和学生参与会议进程。各大学进行作品陈述的次序将通过抽签方式确定。各大学将有60分钟时间陈述其两组作品并接受评委会评审。每组最多进行15分钟的陈述，然后评委进行15分钟的点评。在评审过程中，时间限制将得到严格执行。

奖项
首三名可分别获得奖金新元$15,000、新元$10,000及新元$5,000。

"亚洲垂直城市"国际设计竞赛暨研讨会

Competition Themes
Everyone Needs Fresh Air, 2011；Everyone Ages, 2012；Everyone Harvests, 2013.

Competition Objective
According to the requirements of the theme, the objective of the competition is to seek a holistic solution or a new urban paradigm for a rapidly growing Asian city which also faces the issues of sustainability and quality of life.

Competition Method
For the competition, students of architecture and related disciplines from ten universities were tasked to design one square kilometre of land for 100,000 people, taking into account factors such as density, liveability and sustainability specific to the rapid and exponential growth of urbanism in Asia.

Design Considerations
The design should consider the following issues holistically and integratedly：sustainability (environmental), quality of life (inclusiveness and community), feasibility (buildability, financial and social support), relation to context (place, awareness of conditions, climate and cultural milieu) and technical innovation (technology and techniques).

Team Participation
10 Universities from global are invited. Two proposals from each university will be selected and the University team of 1 faculty member and 2 students will be sponsored to attend the jury session and the symposium. Universities are to identify their teams as "Team A" & "Team B" in all their submissions and documents. The jury members will be invited to deliver a paper at the symposium on the theme in the context of densely urbanized Asian cities. The 2 students will be required to present their respective team designs for the competition to the international jury panel.

Jury Proceeding
The jury proceeding will be an open session, allowing other students and academics to attend as audiences. The drawing of lots will be used to determine the sequence of presentation of each university. Each university with its 2 teams will be allocated 60 minutes for presentation and jury feedback. Each team will be given a maximum of 15 minutes presentation time with 15 minutes for jury comment. Strict time keeping will be enforced.

Prizes
The top three submissions were awarded cash prizes of S$15,000, S$10,000 and S$5,000, respectively.

序言 Preface

寻找未来幸福的阶梯
Towards a Sustainable and Happy Urban Future

在新加坡，亚洲垂直城市国际设计竞赛暨研讨会已经进行到第三届。这项竞赛我们给定的最重要的议题就是城市向上发展，换句话说，这个竞赛的宗旨是为了激发人们对立体城市或者是垂直城市研究的兴趣，能够在互联网时代，激发大家发挥各种想象，面对快速城市化的挑战，深入研究一个最理想的城市发展的规模、尺度、空间密度、产业与生活合理的比例，以及可持续发展城市所需要的空气、水、交通、人口和社会管理方面的诸多议题。

之所以会想到设立和支持这样一项竞赛，主要缘起于在中国的城市化进程中遇到的诸多问题和引发的一些思考。

我在中国从事房地产20多年，深切感受到了中国城市化发展急速的脚步。在快速城市化的过程当中，创造了巨量的GDP，也使很多人住进了以前从来没有想象过的豪宅、大屋。城市化也通过交通，特别是轨道交通的联系，使空间变得非常紧凑，人口的流动也非常频繁。这当然是我们事先预见到和乐见其成的，但不容讳言，快速甚至鲁莽的城市化也给我们带来了很多困扰。

第一个困扰，就是土地资源的浪费。摊大饼的发展模式，使中国的城市化一方面土地非常紧缺，另一方面挤占大量耕地，使城市扁平化无限铺开，也使中国原有的生态环境格局遭到了巨大破坏。

第二个困扰，就是在整个的城市化发展过程当中，房地产成为一个主角，而本身城市发展的经济引擎，也就是产业被忽视，或者说没有时间，没有机会，没有心力，没有愿望能够很快的把产业做起来。城市化就变成了房地产化，房地产化就变成了依赖土地财政，而依赖土地财政便成了政府无限发展新区和"摊大饼"的一个致命冲动。

第三个困扰，就是节能环保。城市无限的蔓延，无效的交通，不必要的移动越来越多，同时汽车数量也越来越多，水和空气变得越来越糟，一些都市特有的疾病越来越严重，使人们的健康受到了极大的损伤。城市化破坏了环境，使人们的生活更加不安。其中，在建筑工地上的能耗，占到我们整个城市发展的新增碳排放的1/3。

第四个困扰，由于这样的城市无限摊大饼带来的功能紊乱，人们的居住、工作、就医、娱乐、教育都变得非常之困难，非常不方便。城市化与原来相比带来了效率，但是没有带来期望的幸福。同时，摊大饼的无限发展方式，给城市的管理带来了非常多的问题，包括我们传统的居委会、街道，还有基层政权的有效管理，都出现了扭曲和变形。凡此种种都归结为一个问题：我们究竟需要一个多大的城市，需要一个多高的城市，需要一个怎么样密度的城市，需要一个怎么样使上班和居住有效结合的城市，需要怎么样建成一个高密度可持续发展的宜居、幸福的城市。最后，怎么样在城市化过程中，让人和人之间有更多的微笑，更多心灵的自由，更多便捷、更多快乐的生活。

正是出于这样一种解决问题的冲动，我自觉有一个使命，就是把公司的业务和城市化的发展紧密结合起来，去承担建设一种可持续宜居城市的责任。

怎么样建设可持续宜居城市？我们特别关注新加坡的经验。在新加坡，由于国土非常的狭小，人口密度非常高，又需要按照一个国家的职能来有效使用这些土地，所以对经济社会发展过程中的土地的使用和空间的使用有非常科学合理的规划，在过往50多年里积累了宝贵的高密度可持续宜居城市发展的成功经验。因此，我们决定依托并借助新加坡的研究力量，和新加坡国立大学设计与环境学院一起，共同举办这样一个全球大学生（研究生）关于未来垂直城市的规划设计竞赛。这项竞赛本身由新加坡国立大学来组织，我们世界未来基金会和万通立体城市公司都给予了一些支持。新加坡国立大学充分运用其自身发展经验和各方面资源、力图使这项竞赛能具有国际最高的水准。这主要表现在评委的选择、议题的设定以及评奖过程的公正、专业和有效。国立大学邀请了全球最富有创建性、最富有经验和最富有理想的一些专家评委，并且着眼于亚洲周边高密度城市发展的需要，先后设定了中国大陆成都的一个地块，还有韩国首尔的龙山地块，以及第三届竞赛的越南的一个地块，分别结合城市化的实践设定相应的竞赛主题。

在成都的竞赛当中，研究的是一个城市多大的尺度和密度是最合理的。到了首尔龙山地块，我们设定密度同样的情况下，怎么样解决养老宜居和就业发展的问题。第三届，在越南我们又换了一个主题，就是在密度同样的情况下，怎样通过都市农业的发展，来使垂直城市变得更加可持续，更加具有生气和活力。未来，我们仍然会坚持这样的路线，也就是根据亚洲不同城市的垂直城市的模型和高密度发展的需要，来设定不同的议题，从而使关于垂直城市的研究本身成为一项长期的可持续的创造价值的研究。

每一次的评奖我都会在现场倾听每一组竞赛选手对于他们作品的认真介绍。在这过程中我大开眼界，让我对整个世界充满了无限的欣喜，因为我看到所有最具有创造力的想象，以及一些细致的勾勒和安排。比如说，曾经有一组美国的选手，他们把垂直城市做成了像树一样的自然生长，每一个建筑体可以挂在一个垂直向上的结构体上。我也看到一组欧洲的选手，他们会把原有城市土地的肌理保存得很好，然后使一个很高密度的城市似乎看上去也并不是那么突兀，而且和环境融合得非常之好。更有想象力的，是在老城市不动的情况下，在老城市的上面，架起一个垂直城市的新城，然后再把老城市底下的人再搬到上面去。这样有效地解决了老城的拆迁和发展问题。

应当说，这样的竞赛是没有止境的，第三届也只是刚刚起步，我们第一步规划了五届，未来可能还会有十届、二十届甚至更多，我们研究的议题也不会停下来。因为只要我们的脚步不停下来，我们的研究和思考就不会停下来。我很期待更多对此议题有兴趣的研究者，包括学生，包括从业者，都能够关注和参与到这项竞赛和研究当中。我也很愿意继续支持这项研究。

令人欣喜的是，我们的这些研究并不是纸上谈兵。这些研究给我的启发很多都已经在中国本土发展立体城市的具体实践当中得到了印证和实施。目前，我们在中国，在西安、成都和温州都在发展这样一个理想的垂直城市或者是立体城市。我们也正是围绕着节约土地、产城一体、节能环保和社会和谐这几个目标，来推动中国本土的立体城市的发展。

更令人欣喜的是，不光我们作为一个企业在实践立体城市这样一个梦想，越来越多的城市的领导者以及相关专业部门的领导者，还有越来越多的中国本地的研究人员都非常支持，并且和我们一起推动这项事业的发展。

未来，我相信立体城市在中国一定会大行其道，成为中国城镇化、城市化发展的一个主流的方式。只有这样，中国城市的发展才能够对国土集约使用，对产业的主导发展，对环境生态的保护以及对社会管理的改革起到一个重要的推动作用。

人生有梦，筑梦踏实。让我们一起再加油，通过一次次的竞赛，使我们的梦想更加灿烂，也使我们的梦想在土地上生根发芽，成为引导未来幸福的一个阶梯。

冯仑 万通控股董事长

This year, the third edition of the Vertical Cities Asia (VCA) International Design Competition and Symposium will take place again in Singapore. The intention of the VCA programme is to address the issue of cities that are expanding vertically. In particular, we want to stimulate research efforts into "three-dimensional" or "vertical" cities. In this Internet Age, we hope to provoke imaginative responses to confront the challenges brought about by rapid urbanization, and to explore the size, scale, density, work-life balance of an ideal city while keeping in mind the topics relevant to sustainable development in terms of air, water, transportation, population, social management, etc.

The many pressing issues that resulted from China's rapid urbanization are a main reason why I established and support the VCA programme. Having been in the real estate business in China for over two decades, I have experienced first-hand how quickly China is urbanizing. An enormous amount of GDP has been generated, and many people now live in homes of unimaginable luxury. Transport systems in cities, especially the railways, provide greater connectivity. With increased mobility, urban space also becomes more compact. These are trends that were anticipated and are welcomed. However, there is also no denying that rapid – even reckless – urbanization has raised many complex issues we need to seriously consider.

The first issue is the waste of our land resources. The "pie pattern" of urban development in China creates, on one hand, the scarcity of urbanizable land, and on the other hand consumes large areas of arable land as cities sprawl horizontally in a way that causes great damage to the ecology and environment.

The second issue is the dominant role that the real estate sector has come to play in the whole urbanization process in China. As an economic engine, urban development has overshadowed industrial development. For many Chinese cities, it seems that there is no time, opportunity, willingness or the motivation to develop industries. Urbanization is equated to real estate development, which cities increasingly rely on for revenue. This dependence on land revenue becomes the fatal impulse that drives the government to recklessly develop new districts and expand the city horizontally.

Thirdly, we face the conundrum of conserving our energy resources. Unlimited urban sprawl, inefficient traffic systems that cause an increase in unnecessary trips have led to the rise in the number of automobiles and the worsening air and water quality. Some serious diseases that become widespread in the urban environment are endangering the health of the residents. The process of urbanization has so ruined the environment that people's lives have become more unstable. One clear statistic is that energy consumption on construction sites now accounts for a third of the newly-added carbon emissions due to urban development.

The fourth issue we face is the urban dysfunction brought about by the limitless "pie model" urban sprawl. Residents face difficulties in living, working, healthcare, entertainment and education. While urbanization has improved efficiency, it has failed to bring the expected happiness for the citizens. At the same time, the sprawling "pie model" urban development imposes many problems for the management of the city. The once-effective management undertaken by the traditional neighbourhood and community administration has been distorted.

These complex issues really boil down to these key questions: How big should our cities be? How tall and how dense should we allow our cities to become? How can city residents balance working and living? How can we build a sustainable, liveable and happy city? Last but not least, what sort of urbanization process would promote greater social harmony and freedom for the mind and soul, more convenient facilities and happier lives?

Motivated to seek a solution to these issues, I aligned the mission of my business with the urbanization process, and to take on the responsibility of building sustainable and liveable cities.

We found Singapore's experience to be particularly relevant. The country is densely populated and has very limited land. To use land efficiently, Singapore has had to plan the land use and spatial distribution for social and economical development in a careful and rational manner. Over the last fifty years, Singapore has accumulated valuable and successful experiences in developing a high density sustainable and liveable city. We decided to partner with the National University of Singapore (NUS) School of Design and Environment to launch an international design competition for university students (graduates) on the planning of vertical cities of the future. The competition is organized by NUS and supported by World Future Foundation and Beijing Vantone Citylogic Investment Corporation.

With its academic experiences and connections with global intellectuals, NUS has hosted a programme of the highest international standard. This is reflected in the professional and impartial approach to the selection of the jury member, the competition agenda and themes, and award process. Creative, experienced urban design and planning experts were invited as jury members; and to focus on the needs for high density development in Asian cities, sites in Chengdu (China), Seoul (South Korea) and Hanoi (Vietnam) were identified as the design sites for the first three editions of the competition, along with a unique design theme based on an appropriate aspect of Asian urbanization.

For the site in Chengdu, the focus of the competition was the appropriate size, density and air quality for a city. In the case of the site in Seoul, the question was the balance of elderly care, employment and local development given the same density. For the third edition of the competition, the theme is how to make the city more sustainable, lively and viable by promoting urban agriculture, again given the same density. For the remaining editions, we will follow the same approach, which is to set the design theme based on the urban models and needs of high-density development of different Asian cities. In this way, the research on vertical cities will itself be a long-term, sustainable and valuable discipline.

I enjoyed listening to the presentations of each design team. It was an eye-opening occasion because of the most creative imaginations and the detailed descriptions and planning from around world that I saw. For example, there was a tree-shaped design scheme by a team from the US in which the buildings hang from an upward-growing vertical structure. I recall a European team whose design preserves the original urban texture so well that the high-density city looks quite natural and integrates perfectly with the surrounding environment. Even more creative was the proposal to build a new vertical city on top of the old one, and then move the residents from the old one to the new one on top, thus effectively solving the relocation and development problems for the old city.

There will be more VCA competitions in future: the third edition is already underway. We planned for an initial run of five annual editions, but there could be ten or even twenty more in the future. Research on vertical cities will also continue. As long as there is urban development, there will also be reflection and research. I look forward to more researchers, as well as students and practitioners to undertake research in the topic and to participate in the competition. I would be glad to continue to support the programme.

It is gratifying to note that the research is not remained only on paper. The ideas produced by the VCA programme have inspired us to incorporate them into the development of vertical cities in China. We are currently working on developing the "ideal vertical cities" in Xi'an, Chengdu and Wenzhou. Through promoting the development of Chinese vertical cities, our goal is to conserve land, integrate industries in the city, save energy, preserve the environment, and to build harmonious society.

What is even more gratifying is that we are not the only ones seeking to realize the vision of the vertical city in China. More and more Chinese government officials and relevant authorities as well as researchers are supporting and advocating the vertical city enterprise.

I believe that in the future, vertical cities will be dominant and become the mainstream development model in Chinese urbanization. It is the only solution to achieve intensive land use, promote industry-oriented development, protect the ecology and environment and support social management reform.

In life, it is important to have dreams, and to work diligently to realize them. Let us try our best to make our vision shine brighter with each competition, and to have our dreams take root and become our pathway to the future happiness.

Dr. Feng Lun
Chairman, Vantone Holdings Co., Ltd.

序言 Preface

在城市化的进程中,许多亚洲城市将选择高密度高容量垂直发展的方案,以限制城市过渡扩张、保护农业用地、优化资源配置。这些问题在人口众多、土地资源有限——在亚洲尤其普遍。香港、新加坡即是采取高密度高容量垂直发展方式的典型城市。这种模式的可贵之处在于,高密度将为城市及居民创造更加包罗万象的社区,降低碳及生态足迹;通过提倡多样性、加强连接性和鼓励创造性以激发经济活力。通过创新的设计手段,高层建筑将具备混合功能及社区功能,达到更高的能耗效率,同时降低单位面积的生态足迹指数。

当然,高密度发展必然带来技术和社会的双重挑战。为了解决交通及人口压力、能源及水资源利用、乃至污染、噪声和疾病传播等问题,必须有适当的管理机制、规划设计及研发。密度越高,挑战越严峻。

创办"亚洲垂直城市"国际设计竞赛暨研讨会的初衷,是希望为来自3大洲、10所大学的——香港中文大学、新加坡国立大学、同济大学、清华大学、东京大学、代尔夫特理工大学、苏黎世联邦理工学院、加州大学伯克利分校、密歇根大学和宾夕法尼亚大学——教师及学生提供一个平台,共同探讨城市高密度高容量垂直发展方式带来的机遇与挑战,集合国际智慧直面亚洲的高速高密度城市化,为高密度/容量的紧缩城市提供可行的可持续发展解决方案。

第一届大赛暨论坛的主题为"每个人都需要新鲜空气",本届则通过"每个人都会变老"的主题,关注许多国家正在面临的人口老龄化问题。未来50年,整个亚洲65岁及以上的人口将有显著增加。据估计,亚洲地区该年龄段人口将增长314%——从2000年的2亿零7百万猛增至2050年的8亿5千7百万。第二届竞赛的各参赛队需要面对这股席卷亚洲的银色浪潮带来的挑战,提出解决方案。竞赛设计场地位于老龄化速度最快的国家之一——首尔。

本次竞赛旨在探索老龄化问题,鼓励积极的老龄化应对方案,解决应如何抓住机遇、在维持现有城市功能水平的同时保障居民的终身福祉。解决方案应关注"积极老龄化"及"居家养老"等概念,为年老及年轻一代设计适当的高密度照护环境,这是社区能否成功应对老龄化的关键因素。

竞赛设计要求——一平方公里一个容纳10万居民,看似非常技术化,而实际上,各参赛队伍的解决方案在很大程度上被其所在地区/国家更广阔的文化背景所影响。

代尔夫特理工大学A队的"城市无限"提出一种策略,在保持城市肌理的同时,实现过去与未来的对话。设计重点着眼于建立指导城市生长的条件和规则,同时发掘并强化原有的城市环境。通过提出一系列定义明确、涵盖范围广泛的政策,"城市无限"方案致力于保留那些让城市生动有趣、令人惊喜、独特的特质,在允许城市抱有自主生命力的同时温和的主导它的生长。

同样来自代尔夫特理工大学的获胜方案"居住一生的城市"旨在确保龙山居民能在此生活一生。方案关注基本城市服务设施的便利性,因而步行距离成为定位基本生活设施的工具,如杂货店、理发店、社区中心等。并列第一的两个方案胜在包容性。

获得二等奖的东京大学方案,通过整合安全便捷的交通系统,显示了设计团队对他们的设计对象——老年——人有十分敏锐的理解。最后,同济大学的三等奖方案强化了场地,在考虑历史因素的同时注意成本控制。本书中详细呈现了获奖方案和其他众多未获奖方案,展现了解决城市人口老龄化方案的无穷可能性,而人的独立特色和社会性在这些方案中都得以存续。

除了设计竞赛方案,本书也包括了参赛院校教师及大赛评审撰写的研讨会论文。这些论文主要阐释了启发设计方案的概念或内容;所有论文的集和,成就了一个倡导多样化思路、启发活力思辨的论坛。论文中涵盖的设计方法及理念相当广泛。

本书是多年努力的成果,而努力的出发点就是坚信亚洲城市必须直面快速城市化和城市无限扩张带来的危害。我相信"亚洲垂直城市"项目对解决亚洲城市面临的困境是非常有价值的。

借此机会我想感谢慷慨捐助亚洲垂直城市项目的北京万通立体之城投资有限公司(立体之城)、世界未来基金会及基金会主席冯仑博士。我还要感谢立体之城资助本书的出版并组织竞赛成果展览。这些机构无私捐赠,使我们有机会向全世界敲响警钟直面刻不容缓的挑战。他们的持续支持,将帮助我们不断地寻找棘手难题的解决方案。

我也要感谢来自参赛院校的院长同行、学生及执导教师对大赛及研讨会的专注投入。最后,我要对新加坡国立大学设计与环境学院及其下属建筑系的学生、教职工表示由衷谢意,感谢他们为设计竞赛、研讨会、展览、出版及后勤保障等方方面面付出的辛劳。

参赛院校的设计方案展示出的知识水平和专业热情让我颇受鼓舞。优秀的方案说明大赛够吸引参赛的年轻设计师的能力,也证明了大赛在解决世界性难题方面的价值。新一届设计竞赛决赛在即,青年智囊将如何解决主题为"每个人都丰收满盈"的大赛挑战,我拭目以待。

与此同时,我希望读者能够与出版参与者一样,享受本书带来的智慧激荡。

王才强教授 新加坡国立大学设计与环境学院院长

As Asia urbanises, many of its cities will adopt the paradigm of high density and high capacity vertical cities in order to limit urban sprawl, protect agricultural land and optimise resource deployment. This is especially pertinent in highly populated countries with limited developable land – a condition that is prevalent in Asia. Cities such as Hong Kong and Singapore epitomise this approach. The value proposition is that density works for cities and people by creating more inclusive communities, minimising carbon and eco footprints as well as improving economic vibrancy through enhanced diversity, connectivity and creativity. Tall buildings, if innovatively designed, can support mixed-use and inclusive communities, are energy efficient and has smaller eco-footprint per unit of floorspace.

However there are also challenges – from the technological to the social – associated with high densities. Proper governance, planning, design and research are necessary to deal with problems ranging from congestion, energy and water to pollution, noise and disease transmission. These become even more pertinent when densities become highly elevated.

The Vertical Cities Asia Competition and Symposium was created to provide a platform for staff and students from the consortium of ten participating universities across three continents — The Chinese University of Hong Kong, National University of Singapore, Tongji University, Tsinghua University, University of Tokyo, Delft University of Technology, Swiss Federal Institute of Technology (ETH) Zurich, University of California (UC) Berkeley, University of Michigan, and University of Pennsylvania — to research the various opportunities and challenges associated with this paradigm and spearhead an international effort to confront the realities of Asia's fast-pace high-density urbanization, and formulate appropriate sustainable solutions for very high density /high capacity compact cities.

While the 2011 inaugural event focused on the theme of "everyone needs fresh air", the current competition and symposium themed "everyone ages" cast a spotlight on the critical issue of ageing that confronts many countries. All across Asia, the number of people aged 65 and above is expected to grow dramatically over the next 50 years. For the region as a whole, the population in this age group will increase by 314 percent - from 207 million in 2000 to 857 million in 2050*. Changes that occurred over 50 years in the West are being compressed into 20 to 30 years in Asia. It is this pressing concern of rapidly greying Asian populations that the participating teams were challenged to address in this second of five annual competitions. Seoul, located in one of the fastest ageing countries in the world, was selected as the competition site.

The competition explores and addresses this concern by encouraging new positive approaches to ageing society that identify opportunities for maintaining capacities and well-being over the life course. By observing concepts such as 'active ageing' and 'ageing in place', the solutions aim to develop appropriate high density built environments, for both older and younger generations; a crucial element to successful ageing within the community.

Although seemingly technical in nature in terms of brief – housing a hundred thousand people in a square kilometre – the design solutions are largely conditioned by the wider cultural conditions of the countries/regions of the universities from which these proposals emanated.

'The Open Ended City' from Team A of the Delft University of Technology proposes a strategy which retains the contextual fabric of the city, while extending the dialogue between the past and the future. The design primarily focuses on creating conditions and rules that will guide the growth of the city, while uncovering and intensifying what already exists. By clearly defining a broad set of policies, the proposal aims to preserve qualities that make cities interesting and unique places, while allowing the city to take on a life of its own whilst gently guiding its growth.

'Life Time City', the other winning design from the Delft University of Technology aims to ensure that people will be able to live in Yongsan for their entire lives. The proposal observes the necessity of readily accessible basic urban services. Therefore, walking distances have become a tool for locating basic destinations such as the grocery shop, the hair salon, the community centre, and so on. Their proposal is one of inclusivity.

The second-place team from Tokyo University exhibited a keen understanding of the aged for whom they were designing, through the incorporation of a safe and convenient transportation system into their solution. Finally, Tongji University crafted a solution which enriches the site, taking history into consideration whilst developing a cost efficient proposal. These solutions and the many others detailed in this book, demonstrate the myriad ways in which the challenges can be addressed whilst retaining each people's unique cultural sensibilities and make-up.

In addition to the design competition entries from the participating universities, this book also contains reflections and contributions of the faculty and jury members during the accompanying symposium. Although each proposal presented an argument for the ideas and programmes that inspire the design; seen together, they are an argument for a multiplicity of approaches and for generating a vibrant debate. The spectrum of design approaches and breadth of ideas were also evident in the papers presented at the symposium.

This book is the culmination of years' effort that began with the conviction that Asia needs to confront the realities of rapid urbanization and urban sprawl that besiege many of its cities. It is my sincere belief that the Vertical Cities Asia programme is doing invaluable work to address all these issues.

I would like to take this opportunity to express my gratitude to Beijing Vantone Citylogic Investment Corporation (Vantone Citylogic), the World Future Foundation, and its chairman Dr Feng Lun, for generously supporting the Vertical Cities Asia programme. My gratitude also goes to Vantone Citylogic for sponsoring the publication of this book and the organization of the exhibition. It is these acts of generosity that allow us to spread awareness on the pressing challenges facing our world. And it is with their sustained support that we will continue to develop solutions for these pertinent issues.

I must also acknowledge the support of my fellow Deans and Heads of the schools, as well as the student teams and accompanying faculty for engaging fully with the topic and theme. Last but not least, my special thanks to the students, staff and faculty of the School of Design and Environment and its Department of Architecture for their help, in one way or another, in organizing all aspects of the programme: design competition, symposium, exhibition, publication and the logistics.

Looking at the solutions, I am encouraged by the knowledge and passion demonstrated by the students. Their designs are a testament to the impressive level of young talent gathered at the Vertical Cities Asia competition, and offer tangible proof of the programme's value in addressing the pressing issues facing the world we live in. With such bright minds tackling these challenges, I find myself eagerly awaiting the upcoming competition's results addressing the theme "everyone harvests".

In the meantime, I hope you will find the contents of this book to be as enjoyable and intellectually stimulating as it was for everyone who put it together.

Prof. Heng Chye Kiang
Dean, School of Design and Environment
National University of Singapore

Vertical Cities Asia International Co[mpetition]

2011 — 2015 "亚洲垂直城市"国际设计竞赛暨研讨会

如今，亚洲正在经历着激动人心的快速城市化进程，一场由农村向城市的大规模移民正持续不休地进行着。与此同时，亚洲城市变得越来越密集，越来越高，其范围和强度可谓是史无前例的。现有的城市建筑与规划模式难以容纳增加的人口，如果继续使用这种模式将会对土地、基础设施和环境造成毁灭性的影响。面对这一严峻的事实，新加坡国立大学与世界未来基金会于2011年联合创办了名为"亚洲垂直城市"的国际设计竞赛及研讨会，旨在为世界各地的科研学者和学生搭建一个独一无二的平台，鼓励他们针对全新的未来城市模式展开深入的探讨和研究，集思广益，博采众长，积极应对亚洲飞速发展的、高密度城市化现状，同时也为亚洲地区高人口密度的超级大城市制订行之有效的可持续发展方案，在不影响生活素质的同时，解决人口剧增的问题。

"亚洲垂直城市"国际设计竞奖赛及研讨会由新加坡国立大学设计与环境学院主办，世界未来基金会和北京万通立体之城投资有限公司联合赞助。大奖赛自2011年1月1日启动，每年举行一次，连续持续5年。每一届主办方将根据特定主题所涉及的针对性问题并结合亚洲本身具有的异质性文化选取亚洲某一城市的代表区域作为本年度的竞赛研究对象，邀请全球十所顶尖大学（新加坡国立大学、荷兰代尔夫特理工大学、苏黎世联邦理工学院、香港中文大学、东京大学、同济大学、清华大学、加州大学伯克利分校、密歇根大学及宾夕法尼亚大学）的建筑系及相关院系的学生参赛。参赛者需在一平方公里的土地为100 000人口进行规划设计，同时以亚洲城市急剧发展为前提，考虑密度、生活质量及可持续发展等要素。每所大学均挑选最佳的两个方案参加决赛。主办方希望通过这样一系列世界各地学生之间的竞赛，激发他们对当前城市化背景下重大课题的思考并提出对应解决方案。最终，设计评委将基于可持续发展、生活质量、可行性、切题度及科技创新五大标准评选出获奖作品，并为获奖者颁发奖金。

Nowadays, the devastating effects on land, infrastructure, and the environment caused by rapidly expanding Asian cities are becoming serious and there is a pressing need to find solutions to address the problems

ns...petition & Symposium

of urban sprawl, congestion and pollution faced by Asia's overcrowded cities. Facing this severe reality, the National University of Singapore and the World Future Foundation launched the Vertical Cities Asia International Design Competition & Symposium, which is aimed at setting up a unique stage for personnel engaged in scientific research and studentsall over the world. The competition encourages them to carry out in-depth discussion and exploration into fresh models for future cities, allowing them to benefit from brainstorming and mutual learning as well as respond positively to the current situation of Asia's rapid, high density urbanization while also making practicable and effective sustainable development plans for Asia'sdensely populated metropolises.

The Vertical Cities Asia International Design Competition & Symposium is hosted by The National University of Singapore's School of Design and Environment and jointly sponsored by the World Future Foundation Ltd and Beijing Vantone Citylogic Investment Corp. The competition, which was launched on January 1, 2011, is held annually and has been going on for five years in a row. Each year, the hosts will choose a representative area in Asian cities as an object of study for the annual competition according to specific problems. For the competition, students of architecture and related disciplines from 10 universities were tasked to design one square kilometre of land for 100,000 people, taking into account factors such as density, liveability and sustainability specific to the rapid and exponential growth of urbanism in Asia. Two proposals from each university were selected for the finals. The Design Jury assessed the entries in five areas: sustainability (environmental), quality of life (inclusiveness and community), feasibility (buildability, financial and social support), relation to context (place, awareness of conditions, climate and cultural milieu) and technical innovation (technology and techniques). The top three submissions were awarded cash prizes respectively.

加州大学伯克利分校
University of California, Berkeley

密歇根大学
University of Michigan

宾夕法尼亚大学
University of Pennsylvania

荷兰代尔夫特理工大学
Delft University of Technology

苏黎世联邦理工学院
Swiss Federal Institute of Technology Zurich

2012年"亚洲垂直城市"国际竞赛以"每个人都会变老"为主题,设计对象是韩国首尔龙山区的一个一平方公里的区域。亚洲在未来50年内的老龄化步伐将会加快,65岁或以上的人口预计将急剧增长,从2000年的2亿7百万增加到2050年的8亿5700万,增幅达314%。西方国家在50多年间所面对的老龄化挑战,亚洲国家却得在20至30年的时限内面对同一挑战。本次竞赛正是为亚洲的银色浪潮的到来寻求解决方案,竞赛旨在鼓励参赛队伍提出积极的老龄化应对方案,解决应如何抓住机遇、在维持现有城市功能水平的同时保障居民的终身福祉。设计方案考虑"积极养老"及"居家养老"等概念,目的为老人和青年打造合适的生活环境,这是社区应对老龄化的关键因素。

This year's competition explores the theme "Everyone Ages" and the design target place is located in one square kilometre of land in Yongshan district of Seoul . All across Asia, the number of people aged 65 and above is expected to grow dramatically over the next 50 years. For the region as a whole, the population in this age group will increase by 314 percent - from 207 million in 2000 to 857 million in 2050. Changes that occurred over 50 years in the West are being compressed into 20 to 30 years in Asia. Facing the pressing concern of rapidly greying Asian populations which the participating teams were challenged to address, the competition explores it by encouraging new positive approaches to ageing society that identify opportunities for maintaining capacities and well-being over the life course. By observing concepts such as "active ageing" and "ageing in place", the solutions aim to develop appropriate built environments, for both older and younger generations; a crucial element to successful ageing within the community.

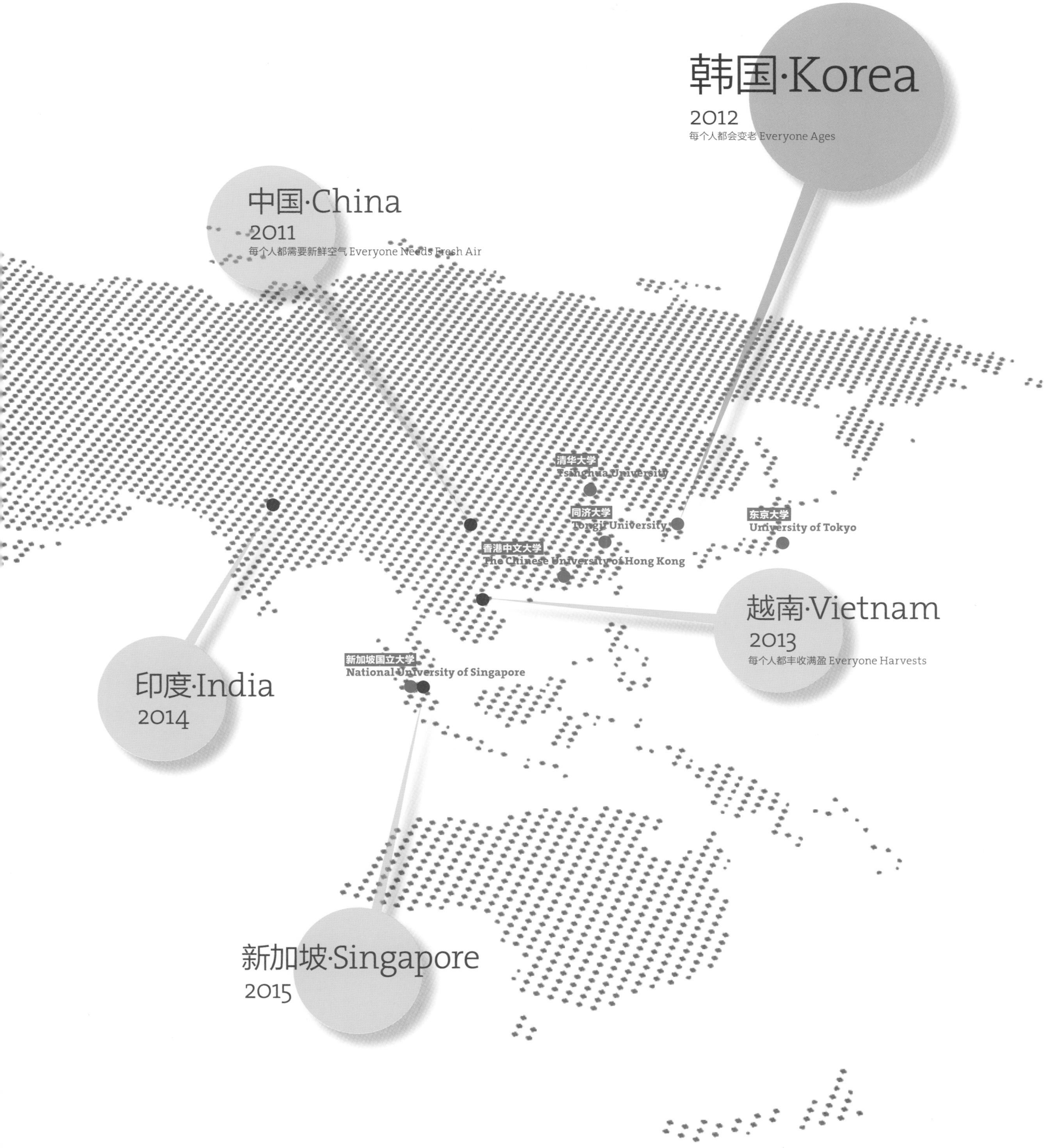

韩国·Korea
2012
每个人都会变老 Everyone Ages

中国·China
2011
每个人都需要新鲜空气 Everyone Needs Fresh Air

清华大学
Tsinghua University

同济大学
Tongji University

东京大学
University of Tokyo

香港中文大学
The Chinese University of Hong Kong

越南·Vietnam
2013
每个人都丰收满盈 Everyone Harvests

印度·India
2014

新加坡国立大学
National University of Singapore

新加坡·Singapore
2015

背景介绍

世界卫生组织称,世界将很快达到一个人口的转折点:2020年以前,65岁和及其以上的人口数量会超过5岁以下儿童的数量。这将是有史以来第一次儿童的数量没有超过他们长辈的数量。世界卫生组织报道,预计65岁和超过65岁的人口数量将从2010年的5.24亿增至2050年的15亿。

纵观亚洲,预计在未来50年内,65岁和65岁以上的人口会显著增加。总体而言,该年龄段的人口会增加314%——从2000年的2.07亿增加到2050年的8.57亿。

有人预测,2050年,亚洲65岁以上的人口将有史以来第一次达到和15岁以下的人口大致相同的数量。西方国家50年发生的变化正被压缩成20-30年在亚洲上演。

亚洲虽然拥有世界上人口最稠密的国家,然而,这片大陆未来的人口正呈现出快速下降的趋势。经济的增长和繁荣已经导致出生率降低,再加上平均寿命的增长,整个亚洲出现了"老龄化社会"的现象。"亚洲四小龙"正快速地跟随着日本人口老龄化的脚步。亚洲人口老龄化这个迫切的问题正是本次参赛队伍需要面临的挑战。

在韩国——2012"亚洲垂直城市"国际竞赛的场地选取地,随着西方医疗卫生体系的引入和平均寿命的相应增长,20世纪早期韩国人口结构就开始出现转变。20世纪60年代中期,由于政府发起了计划生育政策,同时也受经济发展和城市化的影响,韩国人口出生率开始下滑。来自联合国人口署的数据表明,韩国人口的中值年龄已经从1950年的19岁增至2010年的38岁,预计到2050年将会出现52岁的峰值。75岁的平均寿命也在持续增长,从1995年至2000年期间的9.6年增至2010年至2015年期间估计的12年,并预计到2095年至2100年期间将超过17年。与此同时,首尔市的人口增长速度和平均寿命也不断提高。目前,首尔人的平均寿命是80.3岁,到2040年有望达到90岁。同时,低出生率(每个妇女生1.2个孩子)也在加剧韩国人口的老龄化趋势。2008年韩国统计部门统计的数据显示,2012年,首尔65岁以及超过65岁的人口数量预计将超过100万。这个数据也表明,2027年,首尔预计会成为一个超老龄化的社会,65岁以及超过65岁的人口将占总人口的20.3%。

场地选择

本届"亚洲垂直城市"国际设计竞赛的场地位于韩国首尔市龙山区。首尔市是一个拥有1 000万人口的特大城市,约占全国总人口的五分之一。20世纪60年代,首尔市经历了人口的激增。农村地区的贫穷导致大量农民迁往城市,而他们的主要目的地就是首尔。如今,韩国首都圈(SCA)被划分为一个更广阔的区域,包括三个地区:首尔市、仁川市和京畿道,首尔市是其中最大的城市。该首都圈拥有2500万人口,是世界第二大都市区,仅次于日本首都圈。

龙山有着悠久的历史。关于"龙山"作为地名的最早记录存在于一个古老的故事中,这个故事描述了两条龙是如何出现在汉江上的,"龙山"这一名字也由此而来。朝鲜王朝时期(1392-1897),龙山建成了一个码头,商船往来频繁。在韩国的现代历史中,龙山发展成了一个经济活动中心。

龙山区是首尔市的25个区之一。龙山洞(龙山2区)是龙山区的一个行政区,位于首尔市中心,是连接首尔三大中央商务区(江北区、江南区、汝矣岛)的核心地区。此次竞赛的场地占地约一平方公里,毗邻流经首尔中心之后汇入黄海的汉江,距离仁川国际机场50公里左右。该场地也包括龙山国际商务区(IBD)开发区及其周边区域。

龙山国际商务区(IBD)是汉江边上的一个综合开发区项目,从2013年起开始建设。该项目将建成一个国际商务中心、商业区、文化区和居住区,总共包含66幢建筑,其中14幢是摩天大楼。为进行2012年"亚洲垂直城市"竞赛,目前假设龙山国际商务区尚处在前期开发的空置状态,建筑物位于竞赛地点的其他地方。目前,竞赛场地的土地利用主要集中于商业、住宅和基础交通(铁路设施)。参赛者可以选择保留现有的建筑,也可以重新进行设计。

三条地铁线路相交于龙山站,与龙山国际商务区毗邻。此外,地铁中央线穿过比赛场地,与东南部的庆州市相连。河滨北路沿着比赛场地的南端延伸,断开了通往河岸的道路。多项基础设施和公共便利设施项目将作为龙山国际商务区开发的一部分。周边优美的自然景观有汉江、南山和龙山公园。附近还有韩国国立中央博物馆、韩国战争纪念馆、龙山电子市场以及龙山站里的电影城。戍守龙山的美国陆军也驻扎在这片区域,是美国军方在韩的军事总部。

目标愿景

韩国城市(实际上,是世界上所有的城市)如何响应和适应他们的老龄化人口将对人们的生活质量和身体健康产生巨大的影响。在世界卫生组织的定义中,"健康"是指身体上、精神上和社会上的良好状态。世界卫生组织认为,"积极养老"(Active ageing)可以让人们在获得充足的保障和照料的同时参与社会活动,发掘自己在身体、精神、以及社会等方面的潜力。而构建积极养老政策框架的关键目的,就是维护老年人的自主性和独立性。如果人们居住环境的设计能够消除老年人的孤独感,那么强调连续性和识别性的"居家养老"(Ageing in place)也同样是一个积极的方法。

人口稠密的城市为居家养老的概念提供了可能性。2012"亚洲垂直城市"竞赛中关于龙山区的设计,为支持积极养老和居家养老的高密度城市形态指出了两种途径:一种是强调形式的、自上而下的巨型结构途径,另一种则倾向于城市的渐进式成长、自下而上的"软城市化"途径。

这项赛事旨在寻找新颖的设计方案,以迎接即将出现的快速老龄化社会所要面对的挑战,从而维持城市生活的环境平衡状态。本次竞赛旨在鼓励参赛队伍提出积极的老龄化应对方案,解决应如何抓住机遇、在维持现有城市功能水平的同时保障居民的终身福祉。"积极养老"以及"居家养老"等概念为老年人的社会关怀和保障提供了新的途径。这些概念将超越标准社区的范畴而影响设计方法及设计程序,通过一系列措施来调动老年人的积极性,使他们重返社会发挥余热,并为年老及年青一代创建一个适当的照护环境——对于一个让可人们安度晚年的社区来说,这至关重要。

本次竞赛重要的关注点分别是:如何营造一个有影响力的城市形象以传递一个全新的城市生活概念、让中老年人没有受歧视之感;如何倡导"积极养老"的理念,通过建造适宜老年人居住的整体环境来帮助老年人树立独立生活的信心、培养他们的生活自理能力;如何通过包容的、整合的设计来真正实现"居家养老"。"积极养老"和"居家养老"这两种影响城市形态的途径也将逐渐衍生出其它一系列主题:如养老过程化,社会包容性,以及方案的可扩展性等(聚焦于建筑、街区及地区等不同尺度)。

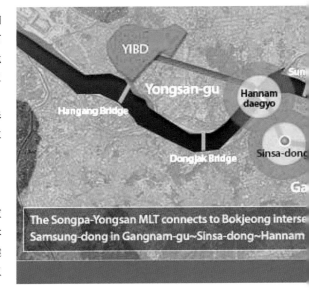

对页及下图:龙山国际商务区由Daniel Libeskind设计总体规划,据预计,龙山国际商务区将于2013年开始建设,并于2016年竣工。开发商为Dreamhub Project Financing Vehicle Co, Ltd.。

Opposite page and below: Yongsan International Business District is designed by Daniel Libeskind. According to the plan, the project will construct in 2013 and complete in 2016. The developer is Dreamhub Project Financing Vehicle Co, Ltd.

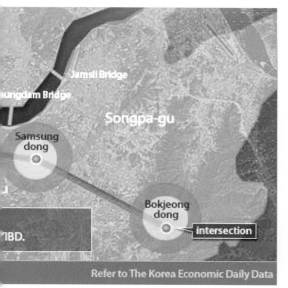
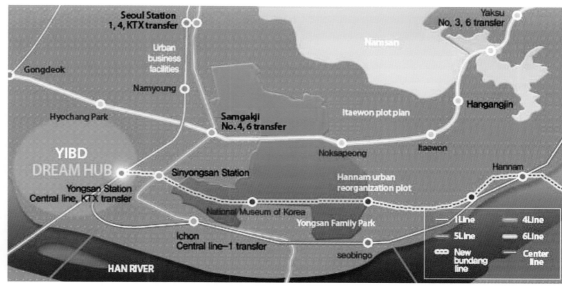

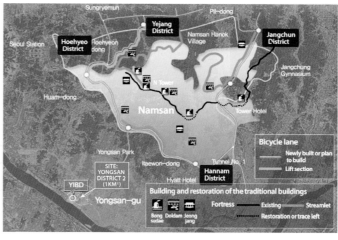
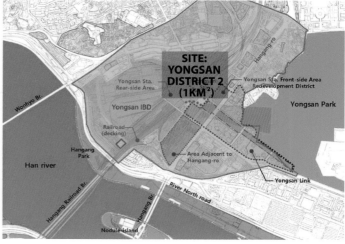

Background Introduction

The world will soon reach a demographic milestone, says the World Health Organization (WHO); before 2020, the number of people aged 65 or older will outnumber children under the age of five years. This will be the first time in recorded history that children have not outnumbered their elders. Reports the WHO, the number of people aged 65 or older is projected to grow from an estimated 524 million in 2010 to nearly 1.5 billion in 2050.

Across Asia, it is expected that the number of people aged 65 and above will grow dramatically over the next 50 years. For the region as a whole, the population in this age group will increase by 314 percent – from 207 million in 2000 to 857 million in 2050. It has beenpredicted that in 2050, for the first time in history, Asia will have approximately as many people over the age of 65 as under the age of 15. Changes that occurred over 50 years in the West are being compressed into 20 to 30 years in Asia.

Asia contains some of the world's most populous countries. However, the demographic future of the continent is fast becoming one of population decline. Economic growth and prosperity has led to both low birth rates that, coupled with increased life expectancy, are generating 'senior societies' across Asia. The 'Asian tigers' are fast following Japan's greying population profile. It is this pressing concern of rapidly greying Asian populations that the participating teams are challenged to address.

In South Korea, where the Vertical Cities Asia 2012 competition site is located, demographic transition began in the early twentieth century with the introduction of Western medical and health systems and an associated increase in life expectancy. The fertility rate began to decline in the mid 1960s with a government-initiated family-planning programme, being influenced also by economic development and urbanisation. Data from the United Nations' Population Division reveals that the median age of the country's population increased from 19 years in 1950 to 38 years in 2010, and is expected to peak at 52 years in 2050. Life expectancy at age 75 (both sexes) is increasing from 9.6 years in the period 1995–2000 to an estimated 12 years in the period 2010–2015, and has been predicted to extend beyond 17 years in the period 2095–2100.

Seoul has seen accelerating population growth with increasing life expectancy, which is currently 80.3 years and is expected to reach 90 years by 2040. Simultaneously, a low birth rate (1.2 children per woman) is contributing to an ageing trend for Seoul's population.Data gathered by Statistics Korea in 2008 indicated that the number of people aged 65 and older in Seoul was expected to surpass one million in 2012. The data also suggested that in 2027, Seoul is expected to become an ultra-aging society with the number of people aged 65 and over accounting for 20.3 percent of the population.

Location Selecting

The site for the 2012 competition is the Yongsan-gu ('Yongsan') district of Seoul, Republic of Korea. Seoul is a 'megacity' with a population of over ten million – around one-fifth of the population of the country. Seoul experienced a surge in population in the 1960s; poverty in rural areas led to the mass migration of farmers to the cities, and the primary destination was Seoul. The 'Seoul Capital Area' (SCA) is now classified as a much broader area containing three districts: Seoul, Incheon, and Gyeonggi-do. Seoul is the largest of these. The SCA has a population of over 25 million, and is ranked as the world's second largest metropolitan area after the "Greater Tokyo Area".

Yongsan has an extensive history. The oldest record of 'Yongsan' as the name of a place is in an ancient story describing how two dragons appeared over the Han River. 'Yongsan' means 'dragon mountain'. During the Chosun Dynasty (1392–1897), Yongsan accommodated a wharf frequented by cargo and merchant ships. In Korea's modern history, Yongsan developed as a hub for economic activities.

Yongsan-gu is one of Seoul's 25 districts. Yongsan-dong (Yongsan District 2) is a ward of Yongsan-gu, and is located in the centre of Seoul. It is a core district that connects the city's three main central business districts (Gangbuk, Gangnam, and Yeouido). The competition site (sized approximately 1 km^2) is adjacent to the Han River, which runs from the middle of Seoul to the Yellow Sea. Incheon International Airport is around 50 kilometres away. The site consists of the Yongsan International Business District (IBD) development area and surrounding zones.

Yongsan IBD is a proposed mixed-use development beside the Han River. Construction of the IBD will begin in 2013. The project will create an international business district as well as commercial, cultural, and residential areas. It will incorporate 66 buildings including 14 skyscrapers. For the purposes of the 2012 Vertical Cities Asia competition, the Yansong IBD site is being considered in its vacant pre-development state. Buildings exist elsewhere on the competition site. The current land use of the competition site is commercial, residential, and transport (railway facilities). Entrants have the option of retaining the existing buildings or designing on a tabula rasa.

Three subway lines intersect at Yongsan Station, which is adjacent to Yongsan IBD. In addition, the Jungang Line (or Central Line) runs through the site and connects with Gyeongju in the southeast of the country. Riverside North Road runs along southern end of the site, interrupting access to the riverside. Multiple infrastructure and public amenity projects are planned as part of the Yongsan IBD development. Adjacent natural and scenic environments include the Han River, Nam Mountain (Namsan), and Yongsan Park. Also in proximity are the National Museum of Korea, the War Memorial of Korea, Yongsan Electronics Market, and a cinema multiplex in Yongsan Station. United States Army Garrison Yongsan is also located in the area; it serves as the headquarters for the U.S. military presence in South Korea.

Target & Vision

How Korean cities – and indeed, cities the world over – respond to and accommodate their ageing populations will have an enormous impact on quality of life, and will significantly affect health. In the WHO's terms, 'health' refers to physical, mental, and social well being. 'Active ageing', says the WHO, allows people to realise their potential for physical, social, and mental well-being throughout the life course and to participate in society, while they are provided with adequate protection, security, and care when it is needed. Maintaining autonomy and independence for older people, says the WHO, is a key goal in the policy framework for active ageing. Ageing in place, it follows – with its emphasis on continuity and identity – can be a positive approach if living environments have been designed to combat isolation.

Dense cities bring particular viability to the notion of ageing in place. The proposals for Yongsan in the Vertical Cities Asia 2012 competition indicate two approaches to the physical form of a dense city that supports active ageing and ageing in place: a top-down mega-structure approach that emphasises form, and a bottom-up "soft urbanism" approach that favours the incremental growth of the city.

The competition seeks innovative design solutions for a balanced environment for urban life addressing and anticipating the challenges of a rapidly ageing society. It encourages new positive approaches to ageing society that identify opportunities for maintaining capacities and well-being over the life course. Concepts such as 'active ageing' and 'ageing in place' with new approach to accessibility, social care and support for elderly are expected to affect design solutions and programs which exceed the standard community club repertoire and incorporate a range of opportunities to activate the elderly and bring them back to workforce, and to develop appropriate environments, especially the built environment, for both older and younger generations, which is crucial to successful ageing within the community. Key issues of concern are: how to create an influential imagery and new concepts of living reflecting the de-stigmatized stand on elderly and ageing; how to encourage 'active ageing' and build up the competency and ability of the elderly to stay independent by providing holistic approach to supportive environments; how to allow 'ageing-in-place' through inclusive and integrative design.

Variations of the two approaches to city form appear across the entries, and a number of themes regularly emerge: ageing as a process, social inclusion, and scalable solutions (focussing on buildings, blocks, and districts).

COMPETITION ENTITIES
参赛作品

在2012年亚洲垂直城市大赛中，来自亚洲、欧洲及美国的参赛学生，提出全面的可持续都市规划模式，以舒缓亚洲城市老龄化的问题。其中，来自荷兰代尔夫特理工大学两支参赛队伍巧用创意解决城市化问题，在国际设计比赛中双双摘下桂冠，东京大学的B队和上海同济大学的A队则分别荣获第二和第三名。参赛团队关于首尔龙山区的提案为支持"积极老龄化"和"居家养老"的稠密城市的实质形态指出了两种途径：一种是强调形式的自上而下巨型结构途径，另一种是赞同城市渐进式成长的自下而上"软城市化"途径。所有方案都提出了有关这两种城市形态变化的类型，有规律地浮现了许多主题：老龄化过程，社会融合和可扩展的解决方案。

In the Vertical Cities Asia 2012, students from Asia, Europe and the US offer holistic solutions to address issues relating to sustainability and aging population in Asian cities. Two teams from Delft University of Technology win International Design Competition with innovative solutions to address urbanisation problems. The second-prize winning scheme from the University of Tokyo, Team B and the third-prize goes to Tongji University, Team A. The proposals for Yongsan in the Vertical Cities Asia 2012 competition indicate two approaches to the physical form of a dense city that supports active ageing and ageing in place: a top-down mega-structure approach that emphasises form, and a bottom-up "soft urbanism" approach that favours the incremental growth of the city. Variations of the two approaches to city form appear across the entries, and a number of themes regularly emerge: ageing as a process, social inclusion, and scalable solutions.

获奖案例：并列一等奖 JOINT FIRST PRIZE

The Open Ended City
TU Delft, Team A

城市无限
荷兰代尔夫特理工大学 A组

导师 Tutors:
Karin LAGLAS,
Henco BEKKERING, Kees KAAN,
Mitesh DIXIT
设计团队 Team members:
Stef BOGAERDS, Samuel LIEW,
Jan Maarten MULDER,
Erjen PRINS, Claudio SACCUCCI

我们在方案中提出一种策略，希望在保持城市真实性的同时，展开现实和未来的对话。
We propose a strategy to retain the authenticity of Yongsan while extending the dialogue between what is existing and what is to come.

我们在方案中提出一种策略，希望在保持城市真实性的同时，展开现实和未来的对话。我们的设计主要关注于创造一定的条件和规则，以引导城市的发展。我们的目标是发现并强化已有事物，城市的环境肌理是我们设计策略的主要驱动力。

这些社区里分散着不同的功能区，而不是只有一种活动的单一社区。社区的核心是混合服务中心，可以为老年人提供医疗保健、学校教育和社区功能。

我们发现，欧洲城市规划中公共广场的原型在韩国是以校园的形式而存在。这是一种很有趣的重复——是综合/混合规划以一种新形式呈现的机会，这种规划建立于韩国社会对教育极端重视的前提之上。新型的公园和滨水区也将会成为有益于户外活动的地方。

与采用宏伟和正式姿态的总体规划不同，我们项目的基础前提是首先保证龙山的长远发展。城市到底是怎样发展的？城市常常因为人口在某一区域的流动而收缩或扩展，而人口的流动又归因于该地区的规划。发展的规模和发展后的城市条件是由于市政府在该地区的政策和法规而产生的。

通过清楚地定义一套广泛的策略，我们的目标是在保留那些让城市变得有趣，令人惊喜的独特物质，在允许城市保有自主生命力的同时，温和地引导它的生长。

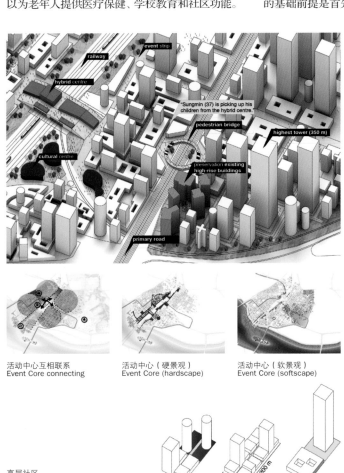

活动中心互相联系
Event Core connecting

活动中心（硬景观）
Event Core (hardscape)

活动中心（软景观）
Event Core (softscape)

高层社区
Highrise Community

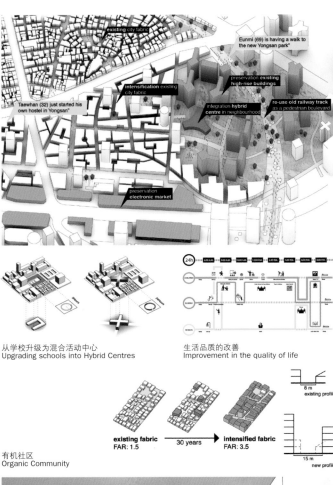

从学校升级为混合活动中心
Upgrading schools into Hybrid Centres

生活品质的改善
Improvement in the quality of life

有机社区
Organic Community

We propose a strategy that is able to retain the authenticity of the city while extending the dialogue between what exists and what is to come. The design primarily focuses on creating conditions and rules that will guide the growth of the city. Our aim is to uncover and intensify what already exists. The contextual fabric of the city remains as the main driver for our strategy.

The idea of community is explored through programmatic circles. These are not a formal gesture, but rather a means of organising programmes in a meaningful way that provides for the needs of an ageing population in a walkable 400-metre radius. These communities disperse functions as opposed to creating agglomerations of one kind of activity. At the heart of the communities are the hybrid centres, which cater for elderly healthcare, schooling, and community functions.

We discovered that the archetype of the 'public square' in European urban planning was present in Korea in the form of schoolyards. This was an interesting overlap – an opportunity for a new form of integrated/mixed programme that builds on the high priority that Korean society places on education. The newly formed park and waterfront will also become a place conducive to outdoor activities.

As opposed to adopting a master plan defined by grand and formal gestures, one of the cornerstones of our project is prioritising the long-term growth of Yongsan. How exactly do cities grow? They usually contract or expand as a result of the flow of people in and through a certain area as a result of the programme of that place. The sizes of developments and resulting urban conditions emerge from the policies and rules set in place by municipalities.

By clearly defining a broad set of policies, we aim to preserve qualities that make cities interesting and at times unpredictable and unique places. By doing so, we allow the city to take on a life of its own whilst gently guiding its growth.

Below: These communities disperse functions as opposed to creating agglomerations of one kind of activity. At the heart of the communities are the hybrid centres, which cater for elderly healthcare, schooling, and community functions.

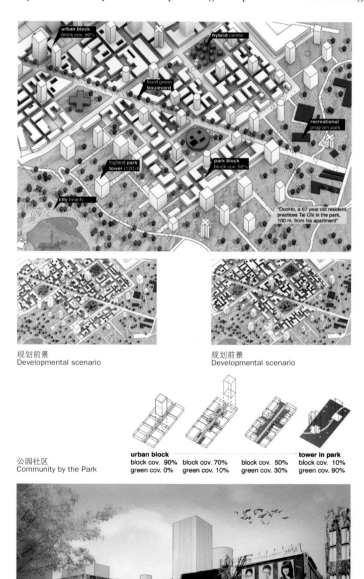

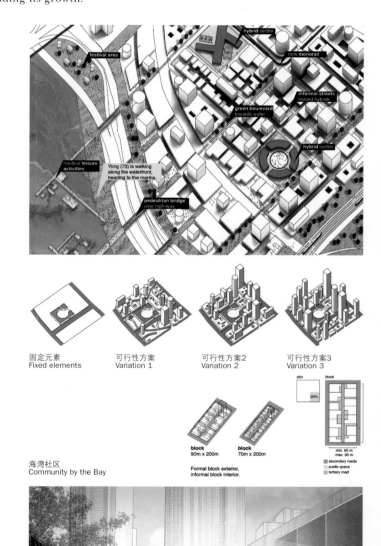

城市无限

  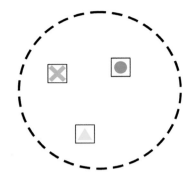 

设计方案
预计到2050年，老年人口比例（38%）和工薪阶层比例（52%）的集合将十分令人担忧。然而与其将老年人视为一种负担，我们不如将其视为一次机会，来解决韩国正在面临的一系列问题。
1. 加强并建设社区
2. 动员妇女进入劳动力市场
3. 使老年人融入社会
4. 组织社区，从而使商店、康乐设施在步行范围内合理分布。

Proposal
By 2050 the convergence between the percentage of elderly (38%) and the working class(52%) is quite alarming. However rather than seeing the elderly as a liability, we see an opportunity to address several issues that S.Korea is currently facing.
1. Strengthen and build communities
2. Mobilise women into the workforce
3. Integrate Elderly into society
4. Organise communities to allow for convenient distribution of shops and ammenities, prioritising walkability.

适宜步行的社区
随着年龄的增长，人们长距离行走的能力逐步减弱。因此，需要优先考虑将功能设施和康乐设施安排在以步行5分钟为半径的区域内，由此人们便不必年老后搬进养老院，而可以在一个社区中居住一生。
我们通过环形功能空间来探索社区理念。这些功能空间并不以正式的姿态存在，而是以一种有意义的方式组织各种功能，来满足老年人群在400米半径步行范围内的需求。

Walkable Communities
As people grow older, their ability to travel long distances diminishes. It is therefore a priority that functions and ammenities be placed within a walkable radius of 5 minutes, so that people are able to continue living in one neighbourhood their whole life without having to move to an elderly home once they get older.
The idea of community is being explored through the programmatic circles which are not a formal gesture, but rather a means to organize programs in a meaningful way which would provide for the needs of an ever ageing population in a walkable 400m radius.

混合中心
混合中心位于社区中心，以满足老年医疗保健、学校教育和社区功能的需求。
这些中心将起到城市再生和发展的催化作用，它们会吸引更多的人在附近区域居住，同时更承担了服务居住者和促进交流的功能。

Hybrid Centres
At the heart of the communities are the hybrid centers which cater for elderly healthcare, schooling and community functions.
They will serve as catalysts for urban regeneration and growth, these centers will attract more people to live in the surrounding areas while serving the people and facilitating exchange.

韩国街区
我们的设计试图通过整合城市街区的高层建筑和低层建筑类型，以及街区内的非正规道路，来增加城市街区的密度。其有利于形成高密度街区以及提高日常的街区生活质量——这是我们在首尔之旅之后，认为对宜居城市最为重要的因素。

The Korean Block
Our design has tackled ways in which to densify the urban block, integrating highrise and lowrise typologies as well as informal routing through the blocks. This allows for high densities and the informal qualities of the street life that we recognised as being vital for a livable city after our experience in Seoul.

The Open Ended City

社区内功能分区
Programmatic distribution within communities

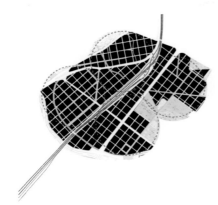

现有道路 vs 新规划的次级道路
Existing vs. New Secondary roads

根据环境调整第三级道路
Tertiary Roads adapting to context

置于社区关键节点的混合中心
Hybrids placed at key nodes with communities

与采用宏伟和正式姿态的总体规划不同,我们项目的基础前提是首先保证龙山的长远发展。
As opposed to having a masterplan defined by grand and formal gestures, one of the cornerstones of our project is prioritizing the long term growth of Yongsan in the design process.

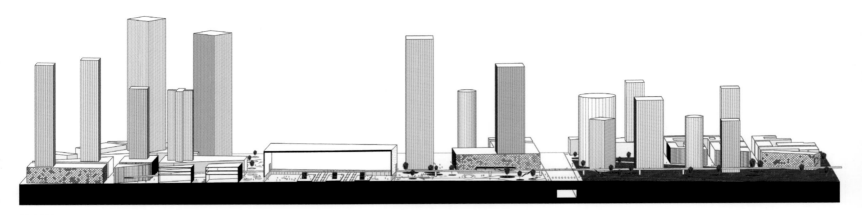

剖透视图
Exploded View

剖透视图
Exploded View

获奖案例：并列一等奖 JOINT FIRST PRIZE

Lifetime City
TU Delft, Team B

居住一生的城市
荷兰代尔夫特理工大学 B组

导师 Tutors:
Karin LAGLAS,
Henco BEKKERING, Kees KAAN,
Mitesh DIXIT

设计团队 Team members:
Laura DINKLA, Katerina SALONIKIDI,
Maria STAMATI, Johnny TASCON,
QIU Ye

龙山——一个适合所有年龄阶段的人居住、生活并融入社区的城市。

Yongsan – a city where everyone in every age can find a place to settle, live and be part of an ongoing process of the community.

```
90 000 – 105 000
75 000 – 90 000
60 000 – 75 000
45 000 – 60 000
30 000 – 45 000
15 000 – 30 000
10 000 – 15 000
< 10 000
```

下图：密度。"亚洲垂直城市"竞赛方案
总面积：2.3 km²
人口：230 000
建成区面积：13,9 m²
FAR: 6

Below: Density. VCA Competition scenario
Area: 2.3 km²
People: 230 000
Built area: 13,9 m²
FAR: 6

走在首尔市区的大街上，你会发现周围的人们摩肩接踵，川流不息。他们吃着小吃，喝着饮料，到处充满了欢声笑语。他们会走进商店闲逛，或者坐在公园里的长凳上。他们在这个城市生活、变老，而这个城市也在发展。不能简单的说每个人都在变老；确切的说，一切都在变老。我们的设计要确保这种情况在将来仍是可能发生的。人们将可以在龙山度过他们的一生，龙山也成为首尔的一个"居住一生的城市"。

这座值得居住一生的城市将会面临一些挑战。随着韩国在二十世纪的繁荣和发展，其经济和人口增长迅速。在这个发展过程中，首尔的城市边界并没有向外扩张，但城市变得更加密集了。于此同时，老年人口数量也在增长，而这带来的影响将是巨大的。据估算，在2050年，首尔有50%以上的人将会超过55岁。这些事实都对我们的设计产生了影响。

城市的基本服务功能应该让每个人都能方便获得。因此，合适的步行距离成为了我们在方案中设置"目的地"（destination）所考虑的主要条件，这些"目的地"包括食品杂货店、发廊、社区中心、公园和朋友的家等等。

具有高密度的新龙山并不需要消除原有的城市生活，消除那些经过多年而形成的社区关系。的确，利用高层公寓楼和办公大楼取代邻里社区的情况曾经在首尔出现过。但在我们的方案中，为了应对首尔人口老龄化的挑战，我们提出了构建社区城市的想法。龙山，不同年龄的每个人都可以在这里居住、生活并融入社区。

Walking through the city of Seoul, you find yourself surrounded by people. They eat, drink, and laugh. They walk into shops or sit on a bench in a park. They live and they grow old in a city that lives and grows. It's not simply the case that everyone ages; rather, everything ages. Our design ensures this is still possible in the future. People will be able to live in Yongsan for their entire lifetime. Yongsan becomes a Lifetime City in Seoul.

This Lifetime City will face some challenges. As Korea flourished in the twentieth century, its economy and population grew. During this growth, the borders of Seoul didn't expand outwards. Instead, the city densified. At the same time, the elderly population has grown. The effect is enormous. Estimates suggest that in 2050, more than 50% of people will be 55 years of age or older. This fact influences our design.

Basic urban services need to be accessible by everyone. Therefore, walking distances have become a tool for locating basic destinations such as the grocery shop, the hair salon, the community centre, the station, the park, a friend's home, and so on.

The new high density in Yongsan does not need to be the result of a process of wiping out urban life that has evolved through the years in the form of neighbourhoods. Indeed, the replacement of neighbourhoods with high-rise serialised apartment and office complexes has happened before in Seoul. In order to face the social challenges of ageing in Seoul, we propose a city of neighbourhoods. Yongsan: a city where everyone of every age can find a place to settle, live, and be part of the ongoing processes of the community.

右图：Dewon-dong，一个韩国人的生活圈。**下图**：首尔人口结构示意图

Right: Safe world of a person in Seoul, Dewon-dong **Below:** Diagram of Seoul population structure

1. Kim女士（71岁）的家
2. Kim女士的邻居家
3. Kim女士的亲戚家
4. Kim女士的美发沙龙
5. Kim女士的教堂
6. Kim女士的超市
7. Kim女士的电车站
8. 距Kim女士家最近的公园

1. Mrs Kim's house (female 71 years old)
2. Mrs Kim's neighbor
3. Mrs Kim's relative
4. Mrs Kim's hair-salon
5. Mrs Kim's church
6. Mrs Kim's supermarket
7. Mrs Kim's metro station
8. Mrs Kim's closest park

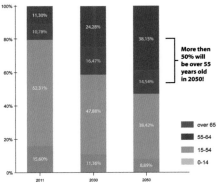

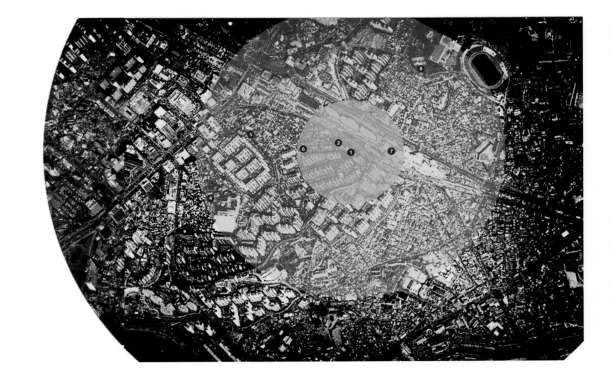

电子市场—公园 Electronic Market - Park

随着数字技术的发展,首尔的电子市场在最近几十年迅速发展,并从城市中的一个位置搬移到另一个位置。将来,我们希望将电子市场放置到公园旁,游客可以在购物的同时参与各种文化活动。

Along with digital technology, Seoul's electronic market has grown in time and has been moving from one location to another within the city during recent decades. For the coming time we propose it to be placed next to the park, where visitors mix the shopping experience with cultural activities.

Uk ch'on河流 Uk ch'on Water Stream

随着电子市场搬迁到新的地方,几十年前被覆盖的Uk ch'on河流因此得以显露出来。河流的引入使得该地区重新富有活力,并再次恢复了因电子市场而被破坏的居住氛围。

As the electronic market is moved to its new location, the revealing of the Uk ch'on water stream which was covered several decades ago, takes place. The introduction of water revives the area and brings back the residential atmosphere which had been derogated from the electronic market.

购物中心和中央商务区 Shopping mall and CBD

铁路沿线也正在发生着变化。一座400米高的高楼取代了以前的大型购物中心"I' Park"。

Changes are taking place along the railways. A tower of 400 meters height and its platform take the place of the "I' Park", former shopping mall.

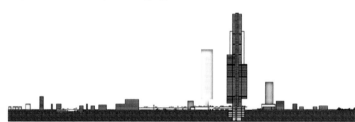

原有社区 Original Neighbourhood

在地段的1号区域是原有社区,其原有的社区特色可以应用到我们新设计的社区中。在我们的方案中,原社区的主干道被保护下来,以保留其地方特色。

The original neighbourhood as identified in district 1 of our site, has characteristicts to be applied in the proposed neighbourhoods. The neighbourhood backbones are protected, preserving their local characteristics.

居住一生的城市
Lifetime City

绿地 Green

在首尔，绿地资源比较缺乏，而取代美国龙山基地的新城市公园（West 8提议）将解决这个问题。我们在场地中设计了东西向的绿带，以让人们方便地进入公园。此外，一条额外的南北向绿带沿着重见天日的Uk ch'on河，将人们带往汉江的北岸。

Seoul lacks green area, a problem to be solved by the new city park (West 8 proposal) which is replacing the U.S. Yongsan base. In order to make the park accessible to people, green connectors crossing our site from west to east are proposed. An extra north-south connector rediscovering the Uk ch'on stream brings people to the north bank of the Hang River.

社区 Neighbourhoods

社区就是首尔的细胞。漫步于龙山，我们可以感受到城市生活的丰富多彩。我们希望通过强化已有社区的特色，并将其应用在新建区域中，从而形成一个由社区组成的城市。垂直社区将同传统社区混合在一起。

The neighbourhood is the cell of Seoul. Wandering around in Yongsan we experience the various characteristics of urban life. We propose a city made out of neighbourhoods, by enhancing the existing ones and recreating its qualities in the areas to be populated. Vertical neighbourhoods will be mixed with the traditional ones.

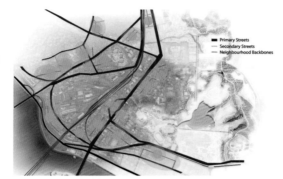

区域 Districts

"特性"是空间最复杂的组成部分之一。图标和标志性建筑有助于理解一个地方的特性。功能、形状、空间、中心性和密度在每个地方构成不同的氛围。首先，我们将整个地段分为9个区域，而每个区域的特性就是社区设计的出发点。

Identity is one of the most complex components of space. Icons and landmarks help in understanding what is the character of a place. Functions, shapes, voids, centralities and densities form different atmospheres in every place. As a first step we splited the site in nine districts. The identity of each one is the starting point for the design of the neighbourhood.

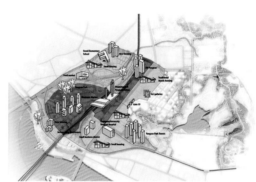

主干道 Backbones

通过了解社区与城市之间的连接元素，我们发现了街道的3种基本类型：a.城市之间的连接，b.城市与社区之间的连接，c.社区内的主干道。那么，我们怎样在场地内定义、加强和复制这些结构及其之间的关系呢？

Understanding the elements that connect the neighbourhoods with the city, we detected three basic types of streets: a. the city connection, b. the access to the neighbourhood and c. the neighbourhood backbones. How could we identify, reinforce and replicate these structures and their relationships within the site?

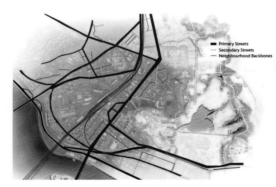
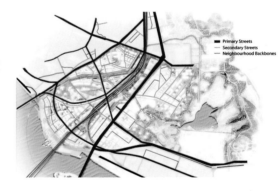

功能 Functions

大多数城市每30年就会被重建一次，然而，成功的功能空间则会一直保留在原来的位置。社区的特性就体现在其内部功能及标志性建筑物，因此，我们可以通过观察现有社区中分布的区域功能来决定方案中新建社区的功能。

Most of the city is rebuilt every 30 years. However, successful functions remain in their original place. Communities acquire their identity from the local functions and the icons within them. From the way local functions are displayed in the existing neighbourhoods we could determine the proposed function mix of each neighbourhood.

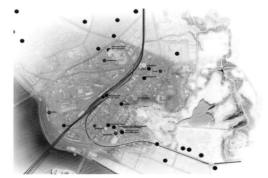

混合利用 Mixed Use

为了使街道富有生机，沿着社区主干道分布着小规模的商业和混合用地。影响力较大的商业和办公用地（CBD，购物中心，电子市场）则坐落于新建和现有的大街两边，迎接着城市中川流不息的人群。

In order to activate the streets, commercial and mixed land use in small scale is introduced along the neighbourhood backbones. High impact commerce and office use (CBD, shopping centre, electronic market) are located along the new and existing main streets, granting flow of visitors from across the city.

获奖案例：二等奖 SECOND PRIZE

Trans-loop
University of Tokyo, Team B

**横向循环
东京大学 B组**

导师 Tutors:
Hidetoshi OHNO, Yusuke OBUCHI, Hiroshi OTA
设计团队 Team members:
Makana YANAI, Taku INAGAKI, Yuya ISHIDA

我们团队的方案是将老龄化解读为生命本身的进程。那么，一个简单的问题便出现了：在一个有着十万人口的城市里会发生什么样的生活事件呢？答案是多种多样的：出生、死亡、入学、毕业、结婚、离婚等等。但是在现代城市中，个人丰富多彩的生活事件会被作为集体经历一起分享和欣赏吗？在首尔，存在着许多问题，比如，妇科医师不足，越来越多的老年人孤独终老。我们必须提出一个新的原型，从根本上改变这个人口过度老龄化的社会。

我们设想创造一个浓缩的"生活事件循环圈"（loop of life events），这里的循环可理解为五种元素的叠加。

第一，正如之前所提及的，它是生活事件的循环：每个人生命过程的可见度得以提高并且连接得以加强。

它也是一个绿色循环：连续且开放的绿地成为城市的基础，在这里，人们可以交流、分享他们的生活。这是一个适于步行的循环：该循环圈周长有3 500米，集合了各种交通工具，如有轨电车和集体捷运。它将老年人基础设施和各种服务项目有效地结合起来。这也是一个天际线的循环：建筑体量同周边环境以及老年人项目的策略性分布形成了呼应，沿着循环圈形成一条明显的天际线，为城市提供了一种独特的视觉特征。

在更大的尺度上，可以说它是首尔这座城市的循环。它将现有的城市绿轴连接起来，成为了一个支撑运输系统的移动中心。这种循环通过分享个人的生活事件并为其提供紧密的支持，提高了人们的老龄化质量。作为一个原型，这一循环结构可以应用于首尔的其它地方，而这些密集的老龄化循环链将改变整个地区。

Our team's proposal began by interpreting ageing as the life process itself. Then, a simple question arose: What kinds of life events are happening inside a city of 100,000 people? The answer is diverse: from birth to death, school entrance to graduation, marriage to divorce.

But is the rich diversity of individual life events shared and appreciated as a collective experience in the modern city? In Seoul, there are many problems such as a lack of gynaecologists and an increasing number of solitary deaths of elderly people. We must come up with a new prototype that will radically change the way we deal with the hyper-ageing society.

Our proposal is to create a condensed "loop of life events". The loop we propose is a superposition of five elements. Firstly, as mentioned, it is a loop of life events where personal life processes take increased visibility and embody enhanced connectivity. It is also a loop of greenery – a continuum of green open spaces, giving the city a backbone where life events will be shared. It is a walkable loop. With a perimeter of 3.5 kilometres, and the integration of various transportation modes such as tram and GRT, it connects ageing-support infrastructure and various programmes efficiently. It is also a loop of skyline. In accordance with the surrounding environment and strategic placement of ageing programmes, the architectural volume along the loop draws a distinct skyline. This will offer a visual identity to the city.

At a larger scale, it is a loop for Seoul. It connects the existing green axis, and becomes a mobile hub to support transportation. This loop will, by sharing the individual's life events and supporting them compactly, improve people's quality of ageing. As a prototype, this loop structure can be applied to many places in Seoul, and the chain of these dense ageing loops will transform the whole region.

TRANS-LOOP

TEAM B
Makana YANAI
Taku INAGAKI
Yuya ISHIDA

WHAT KIND OF "AGING" IS IN A CITY?

BUT, SEOUL IS NOT FULFILLING THE NEEDS OF AGING.....

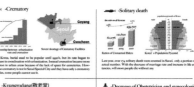

WHAT IF WE CONCENTRATE AGING IN A CITY?

INTRODUCING THE "LOOP"

1. LOOP OF AGING

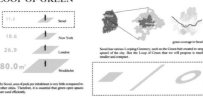

2. LOOP OF GREEN

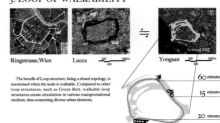

3. LOOP OF WALKABILITY

4. LOOP FOR SEOUL

1. BECOMING GREEN HUB

2. PROVIDE A COMPACT LIVING FOR THE TRANSPORTATION REFUGEE

获奖案例：三等奖 THIRD PRIZE
Soft City
Tongji University, Team A

软城市
同济大学 A组

导师 Tutors:
HUANG Yiru, YAO Dong, ZHU Peidong
设计团队 Team members:
BAO Yinxin, DONG Jia, ZHANG Haibin,
ZHANG Jiawei, ZOU Mingxi, ZHAO Junliang

PRESERVING CITY MEMORY

FRAMEWORK OF VERTICAL LIFE

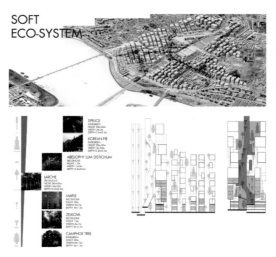

SOFT ECO-SYSTEM

当我们想到"每个人都会变老"这一主题时，我们可能会想到老人如何不能像他们年轻时那样奔跑，或者他们如何不能像年轻时那样轻而易举地搬动东西。我们可能会想到无障碍住宅。

然而，"每个人都会变老"并不仅限于人，它同样可以指一个城市衰老的过程。它指的是人们在一座城市中的一生——出生、学习、恋爱、享受朋友们的陪伴、发生故事、留下回忆，然后讲给子子孙孙们听。这个主题指的是一座住在其中的每个人都会变老的城市。

众所周知，记忆对于在一座城市中老去的每个人来说都很重要。虽然我们是在设计一个新的区域，但我们希望它能体现作为首尔的一个区域的记忆——体现首尔的生活记忆。因此，我们决定前往首尔去寻找答案。当我们回想在首尔遇到的趣事时，一个词浮现在我们的脑海中：软城市。软城市包含了每件事、每个人以及记忆的每个部分。软城市是持续的且便于出入的城市。软城市与自然有着一种崭新的、良好的关系。软城市是一座适宜人们在此逐渐老去的体面的城市。

依据这一理念，我们的区域设计分为三大部分。一部分着眼于如何通过使用地下空间来保留记忆，同时组织抽象的记忆网络。另一部分强调垂直生活框架（交通和公共空间）在规划中的连续性。第三部分展现了如何经由垂直绿化设计和水循环系统来实现与自然的和谐共处。

When we consider the theme "Everyone Ages", we may think about how an older person can't run as they did in their youth, or how they can't lift things as easily as they once could. We may think of barrier-free houses. However, "Everyone Ages" does not only refer to people; it also refers to the process of ageing in a city. It refers to the whole life of people in a city – birth, learning, falling in love, enjoying the company of friends, making memories, and generating stories to tell their children and grandchildren. The theme refers to a city for everyone who will age in it.

As we all know, memory is important for everyone who ages in a city. Although we are designing a new district, we want it to embody memory as a district of Seoul – with memories of a Seoul life. Therefore, we decided to travel to Seoul to find our answer.

When we think about the interesting things we found in Seoul, a word comes to us: soft. Soft City contains everything, everyone, and every part of memory. Soft City is continuous and easy to access. Soft City has a new and good relationship with nature. Soft City is a decent city for every citizen who ages.

According to this, we designed our district with three main parts. One part looks at how to preserve memory using underground space and organising the abstracted grid of memory. Another part looks at continuity through the planning of a framework for vertical life (transportation and public space). The third part looks as how to live with nature via a vertical green plan and a water circulation system.

 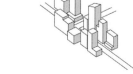 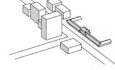   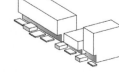 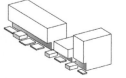 

HILLSIDE — Seoul is located in mountainous terrain. The streets on the slopes make city public spaces closer to the nature and shape the urban life in a soft way.

MODERNITY — Now there are a lot of high-rise residential area in Seoul, people are yearning for the modern lifestyle in a high-rise apartment.

FUSION — The city become a fusion of traditional buildings and modern constructions. The conflict between old and new give more possibilities to the city life.

STREET — The pleasant streets of Seoul are filled with lively public activities. Locals and tourists can both find themselves comfortable and exciting on the street.

MARKET — There are many different types of markets in Seoul, fish market, electric goods market, secondary market and so on. And they are important part of the local's life.

BLOCK — The commercial blocks of Seoul city are with perfect scale and appropriate program of urban life.

获奖案例：优秀奖 HONORABLE MENTION

Soft Yongsan
NUS, Team A

温柔龙山
新加坡国立大学 A组

导师 Tutors:
HENG Chye Kiang, WONG Yunn Chii,
NG Wai Keen, HWANG Yun Hye, Jurgen ROSEMANN,
LOW Boon Liang
设计团队 Team members:
HENG Juit Lian, YEO Wei Ling Diane, PEH Li Lin Stacy

温柔龙山是一座属于每个人的城市。每个人都可以在这座城市中找到自己的人生之路，找到属于自己的生活。在龙山的城市肌理中存在着日常但充满活力的空间，其具有一种有机的混乱，使得各种活动在这里都有可能发生。这是一个年轻人和老年人可以共同工作、娱乐和生活的城市。

我们所设想的"温柔"城市与大都市首尔目前快速发展的情况形成了鲜明对比。这是一座充满活力的城市，居民一生都可以享受富裕而健康的生活方式。

在我们的温柔城市中，"现场演员"是不可缺少的。这些"演员"包括了老年人、白领工作者、蓝领工作者、青少年和儿童。我们在项目方案中采取了设计策略，以创造一种一致的归属感。有了这种归属感，演员们可以相互交流、友好相处，最重要的是他们可以优雅地变老。

为了创建一个温柔且充满活力的龙山，我们采取了三种策略，内在地推动着为居民营造富裕而健康的生活空间。首先，鼓励在高密度城市环境中功能和用途的混合搭配，最大限度地利用空间并促进社会的整合；其次，激活地面、以及地面上空的开放空间；最后，利用通行便利的街道和交通网络来提高整个城市的可步行性和整合程度更高的流动性。

在人口稠密且需求随着时间不断变化的老龄化社会中，温柔城市有着及其重要的意义。通常情况下，因为建成环境和人是分开考虑的，所以通过硬发展形式进行的升级和重建规划是暂时有效，而且花费高，不可持续。因此，建成环境需要用一种更人性化、更灵敏的方法来适应。

"只因并只有当城市是为每个人建造的，它才会具有为每个人提供一切的能力。"——简·雅各布斯

Soft Yongsan is a city for every individual. It allows one to plot a course through the urban fabric, with each individual making Yongsan his or her own place of living daily. There is a reachable space for every intention, with an organic chaos that is attributable to the vibrancy and informal spaces existent in its urban fabric. It is a city where both young and old live, work, and play in adjacency.

We envision a "soft" city that contrasts with existing rapid and hard developments evident in metropolitan Seoul. This is a lively city where dwellers are able to enjoy productive and healthy lifestyles throughout their lives.

Our on-site actors are integral to the softness of the city. They include the elderly, white-collar adults, blue-collar adults, youth, and children. This profiling guides our strategy for creating a unanimous sense of belonging, where the actors can commune, reside, and most importantly age gracefully.

We have adopted three main strategies in order to achieve a soft and vibrant Yongsan, which inherently facilitates productive and healthy living spaces for our dwellers. Foremost, a mix of functions and uses is encouraged within a high-density urban environment to maximise spatial usage and foster social integration. Secondly, open spaces at the ground level as well as above ground are activated. Lastly, a well-connected and accessible street and transport network promotes walkability and better-integrated mobility for all.

A soft city matters in a densely populated ageing society where needs are ever changing along with time. Often, upgrading and renewal projects in the form of hard developments are temporarily effective, expensive, and not sustainable because the built environment and people have been considered separately. Thus, a more humane and sensitive approach to the built environment is required.

"Cities have the capability to provide something for everybody, only because, and only when, they are created by everybody." – Jane Jacobs

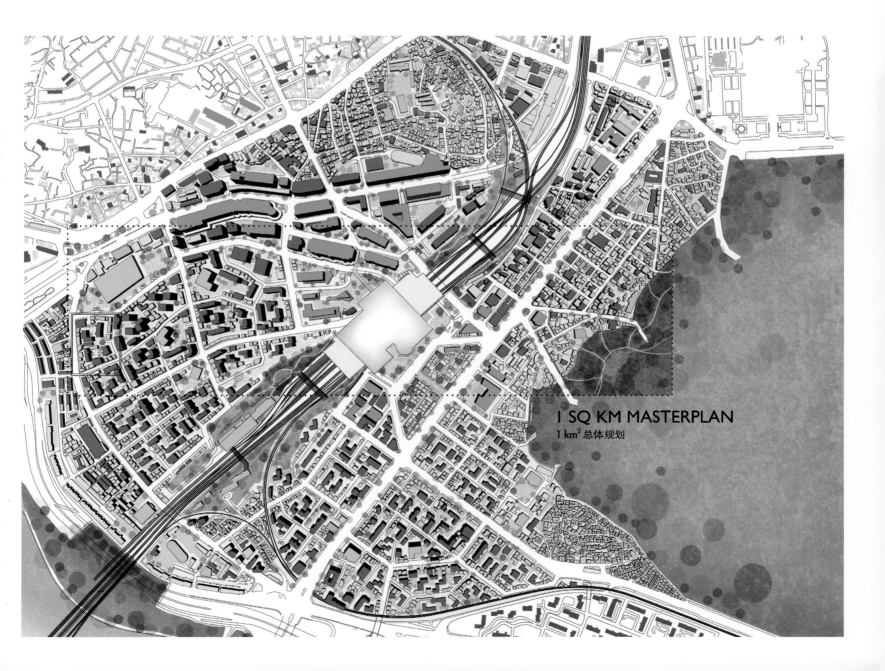

1 SQ KM MASTERPLAN
1 km² 总体规划

获奖案例：优秀奖 HONORABLE MENTION
Ex'Stream
University of Tokyo, Team A
溪流
东京大学 A组

导师 Tutors:
Hidetoshi OHNO, Yusuke OBUCHI, Hiroshi OTA
设计团队 Team members:
Yohei IKAI, Cosmo TAKAHASHI, Shimpei KASAI

我们有关老龄化的思考立足于两个方面。一方面是城市的老龄化问题，另一方面是人口的老龄化问题。此外，考虑到首尔的城市环境，我们的理念在于创造一个适应性强且可以自我生成的城市，同时提高老年人的就业机会——韩国人口结构目前的改变意味着老年人应当继续从事工作。

龙山货场坐落于场地之中，我们保留了场地中原有的铁路轨道。之后，我们增加了运河水道——将河流和城市连接起来，解决了河流和城市分离的问题。

我们城市的每一部分都是由树形的"流群集"（Stream Cluster）组成。该"流群集"有四个优点：第一，每一个群集都可以自给自足，因为其中的集中和分布都是互相共存的；第二，群集的横向和纵向扩张极为便捷，因而非常适合自我发展；第三，"流群集"拥有分形结构，因此能大规模产生多种功能以及就业机会。"流群集"集中的最典型例子便是龙山车站——当到达车站并四处寻找时，你能发现很多的招聘信息。

铁路轨道和运河水道的利用有助于城市的自我发展。建筑材料工厂沿着河流和铁路分布，因而其生产的产品可以方便地运输到这个新兴城市中。新兴城市会逐步地发展起来；即便其建设暂时停止，该城市的集中和分布格局也使得它能很好地运行。

此外，河流和铁路轨道提供了多种多样的风景，也为老年人创造了参加各种活动以及就业的机会。

Our thoughts about ageing focused on two different facets. One is the ageing of the city, and the other is the ageing of people. Moreover, in consideration of the urban environment in Seoul, our concept is to make an adaptable and self-generative city and to increase the job opportunities for elderly people. The changing structure of the population in South Korea suggests that older people should continue working.

The site includes the Yongsan freight yard, and we retained the rail tracks in our new adaptable city. Then, we inserted the canal to solve the problem of the disconnection between the river and the city.

Every part of our city is composed of a tree-shaped 'Stream Cluster'. The 'Stream Clusters' have four beneficial points. The first is that one cluster can support itself, because concentration and distribution coexist in each. The second point is that both vertical and horizontal expansion is easy; thus, this is a reasonable shape for self-generation. The third point is that the 'Stream Cluster' has a fractal structure and, at a large scale, multiple functions and employment opportunities are generated. The most extreme instance of concentration is inside Yongsan Station.

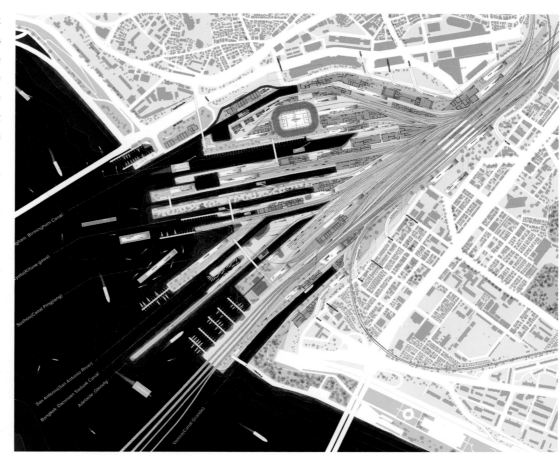

When you arrive at the station and look up, you can see various employments.

Making use of the rail tracks and canals helps with the self-generation of this city. Construction material factories are located along the river and the railways, and its products are easily conveyed to this new city. This new city grows up gradually. Even if its construction is temporarily halted, the city can continue to function well because it has concentration and distribution. Additionally, the water and rail tracks create diverse scenery and activities for elderly people, as well as various jobs.

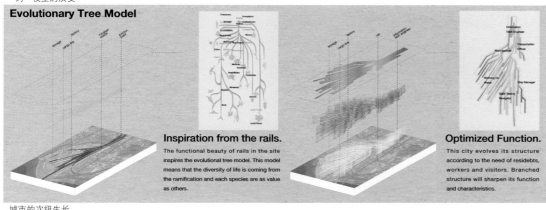

获奖案例：优秀奖 HONORABLE MENTION

Inhabiting the Seams
University of Michigan, Team A

居住于接缝之中
密歇根大学 A组

导师 Tutors:
Monica PONCE DE LEON, Lars GRAEBNER
设计团队 Team members:
Sara ANDERSON, Anna SCHAEFFERKOETTER, Elliot WEISS

在龙山多变的城市结构和自然元素的交叉点上，灵活的建筑类型的植入生成了各个年龄的人群所居住的节点。这些节点连接着城市肌理，并向居民和游客提供社会服务。独特的项目利用每个节点上的自然元素，在城市接缝中产生活动。

四个节点——森林、水、悬崖和农业，各自结合了不同的环境：复原的Uk ch'on、新的和现有的城市结构以及龙山公园。其中，Uk ch'on项目和公园对该区域的发展起着关键性的作用。节点突出了城市结构中近期（与未来潜在）的重要改变，并将其融入到这座城市的生活中。

对于整个场地而言，其策略核心在于这样一种理念：城市的发展，尤其是节点的发展通常并不局限于任何特定的时代或者特定的使用群体。不受时代限制的多用途的功能空间能适应满足所有人的需要。

地段的位置位于现有城市结构之间的接缝之中，即国际商务区（IBD）和龙山公园。地段的边界同样包含了汉江曾经的主要支流——Uk ch'on的历史路径。它曾在快速的城市化浪潮中被填平。随着Uk ch'on生根并成熟，它将作为检测龙山城市结构随时间而发展的指示剂。

每个节点通过对建筑类型的灵活处理，强调城市结构之间的入口连接。典型的类型是一组建筑群，包括具有大型家庭单元和供独身者与夫妻居住的小型家庭单元的中高层建筑，以及能够容纳具备不同自理能力的老年人的辅助低层建筑。建筑类型相互作用，从而创造出一种集中的、活跃的外部空间。

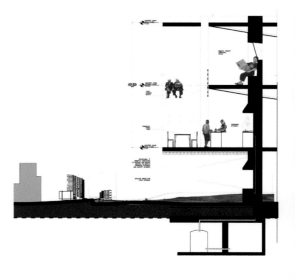

At the intersections of varied city fabrics and naturalistic elements, the insertion of flexible building typologies generates nodes inhabited by people of all ages in Yongsan. The nodes connect the fabrics and provide social services to inhabitants and visitors alike. Unique programmes capitalise on the naturalistic elements present at each node, generating activity at the city seams.

The four nodes – Forest, Water, Cliff, and Agriculture – each relate to a combination of specific conditions: the restored Uk ch'on, new and existing urban fabrics, and Yongsan Park. The Uk ch'on project and the park in particular play critical roles in the development of the site. The nodes highlight these important recent (and potential future) changes in the urban fabric and integrate them into the life of the city.

Central to the site strategy is the belief that the development of the city in general, and the nodes in particular, is not limited to a focus on any one age or user group. Multi-functional programmatic spaces that are not segregated by age can adapt to meet the needs of all people.

The site's location creates seams between the existing urban fabric, the International Business District (IBD) and Yongsan Park. The boundaries of the site also encompass the historic path of the Uk ch'on, once a major tributary of the Han River. It was paved over during a wave of rapid urbanisation. As the Uk ch'on takes root and matures it can serve as an indicator of the passage of time for the evolving Yongsan fabric.

Each node is the manipulation of a flexible building typology, emphasising a gateway link between city fabrics. The prototypical typology is a grouping of buildings that include mid-rises with larger family units and smaller units for singles or couples, as well as assisted living low-rises that can accommodate the elderly at varying levels of dependence. The building types interact with each other to create a focused, active exterior space.

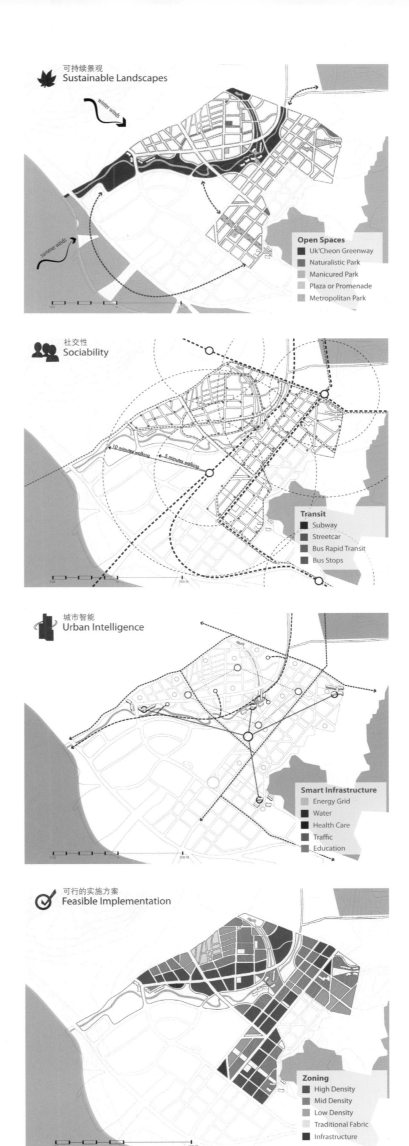

获奖案例：优秀奖 HONORABLE MENTION
Senior Catalyst Cluster
ETH Zürich, Team A

老年人催化群
苏黎世联邦理工学院 A组

导师 Tutors:
Alfredo BRILLEMBOURG,
Hubert KLUMPNER, Kees CHRISTIAANSE,
Michael CONTENTO,
Nicolas KRETSCHMANN
设计团队 Team members:
Carmen BAUMANN

MAIN STRATEGY

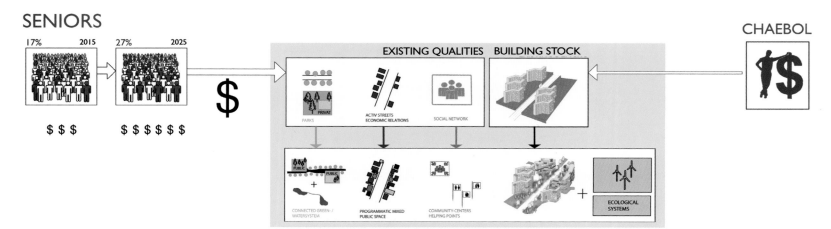

SENIOR-CATALYST-CLUSTER

与现实发展趋势相比较而言，老年人催化群背后的理念是要建成一座不再由单个的/强大的利益相关者决定，而是由集体影响力共同打造的城市。这就是所谓的众包。我们认为这一方法是通过满足基本需求及欲望来提高整体生活水平的最有前途的方法。

在我们的新城市系统中，老年人扮演了主要角色，同时也是催化剂。我们想要通过分配新的任务和功能，使老年人重新成为劳动力群体的一部分，重塑其社会角色，并为他们提供改善自身生活水平的机会，使他们通过利用其自身财力来取得一些改变。

在社会层面之上，我们的城市在经济和生态方面也是可持续发展城市。该方案的精髓在于分析现存的、可用的现场资源，发现他们的潜能，保留现存结构并将其纳入计划中，利用这些资源定义一个保障人人都能享受高质量生活的城市框架。

通过有效的空间管理，利用不同的类型，容纳10万人生活和工作的城市是可以实现的目标：利用场地良好的风力条件，建设水库和污水处理系统以提高生态可持续性。我们的主要概念就是众包。它应是一个战略过程，将会构建一个整合未来发展和项目于一体的框架结构。在我们的计划中，建筑形式并不是中心点。计划的重心集中在发展模式上——循序渐进的场地变革。

我们梦想着不仅能创造出老年人和年轻人之间的协同增效效应，也能创造出大财团经营和私人投资者众包之间的协同增效效应。通过最大化地利用现存资源，涵盖直接关系群体，建设一座社会、人口、规划综合发展的城市。

existing site resources

cooperative building zones

water-treatment system

pedestrian/bicycle network

helping points new activated spaces

neighbourhood meeting points

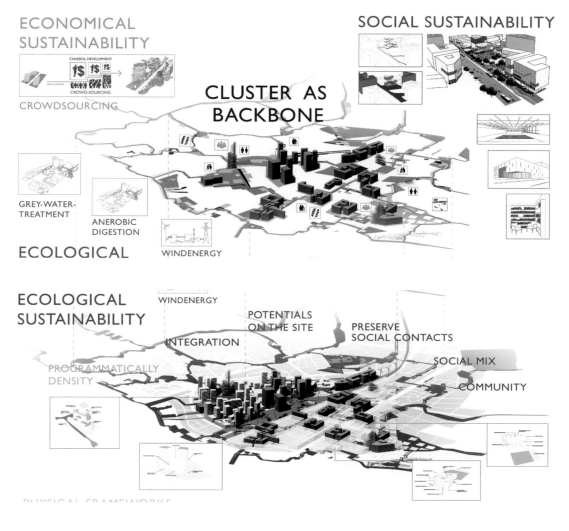

The vision behind Senior Catalyst Cluster is to build a city that — in contrast to the actual tendency — is no longer determined by single, powerful stakeholders but created by the collective influence of many. This is known as crowdsourcing. We consider this approach to be the most promising in order to improve the overall quality of life by satisfying essential needs and desires.

Elderly people are the main actors as well as the catalysts of our new city system. By assigning new tasks and functions, we intend to bring them back to the workforce, reintegrate them into society, and give them the opportunity to improve their living standards. Change shall be carried out through the use of their financial power.

Beyond the social, our city is also sustainable in economic and ecological terms. The essence of the project is to analyse existing, available resources on site, recognise their potential, preserve and include existing structures, and use them to define an urban framework thatguarantees a high quality of life for everyone.

Through efficient space management and the use of different typologies, a density of 100 000 people living and working shall be reached. Ecological sustainability will be improved by utilising the site's good wind conditions and establishing water storage and a grey water treatment system.

Our main concept is the crowdsourcing aspect. It should be a strategic process, generating a framework that can integrate future developments and projects. The architectural form is not the central point of our project. The focus is a model of development —a step-by-step transformation of the site.

The dream is to create synergies not only between elderly and younger people but also between the operations of big chaebols and crowdsourcing with private investors. By devoting as many existing resources as possible and involving directly concerned groups of people, the goal of a socially, demographically, and programmatically mixed city is reached.

STEP-BY-STEP DEVELOPMENT

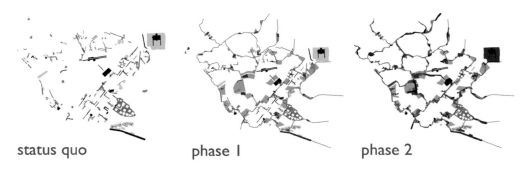

status quo — phase 1 — phase 2

PHYSICAL FRAMEWORKS

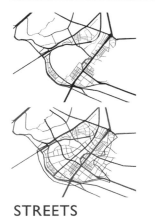

STREETS

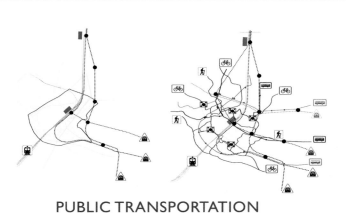

PUBLIC TRANSPORTATION

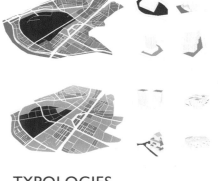

TYPOLOGIES

COMPETITION ENTITIES
其他

Succulent City
University of California, Berkeley, Team A

润泽城市
加利福尼亚大学伯克利分校 A组

导师Tutors: Jennifer WOLCH, René DAVIDS
设计团队Team members: Ekaterina KOSTYUKOVA, Kristen HENDERSON, Aine COUGHLAN

"润泽城市"通过恢复城市与河流之间的联系，在现有城市环境中嵌入一个动态、高效、服务于所有年龄层的网络（network）——其整合了雨水采集、污水过滤、公共娱乐等功能，在阳光、风等自然力量的影响下沿着城市间的生态洼地和商业路线发展。由高楼、地面和空中结构组成的建筑网络与湿地和陆地景观完全融合，并呈阶梯状分布着混合功能，可以为不同年龄层的人们提供服务，同时在不同高度具备移动性和与城市社会之间的关联性。

Revitalising the connection to the river, Succulent City embeds a dynamic and productive network in the existing urban context. It is a network for all ages. Integrating rainwater collection, grey water filtration, and recreational public space into a branching building/landscape system, the network is oriented along bioswales and commercial routes, sculpted by the natural forces on site such as sunlight and wind. The building network of towers, ground, and sky branches is thoroughly integrated with the wet and dry landscape, serving all ages with a gradient of mixed-use programme that benefits those with varying levels of mobility and connection to the urban society.

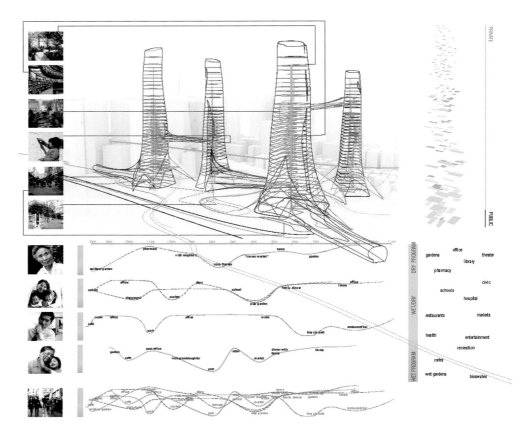

2015
ONE-STORY DENSITY NEIGHBORHOOD

2025
MONORAIL STATION INSIDE VERTICAL CORRIDOR

Toward a Utopia of the Real
University of California, Berkeley, Team B

通往现实的乌托邦
加利福尼亚大学伯克利分校 B组

导师Tutors：Jennifer WOLCH, René DAVIDS
设计团队Team Members: Caitlin ALEV,
Karen GATES, Stathis GEROSTATHOPOULOS

YONGSAN DISTRICT II

MASTER PLAN

SEOUL | SOUTH KOREA
CAITLIN ALEV | KAREN GATES
STATHIS GEROSTATHOPOULOS
COLLEGE OF ENVIRONMENTAL DESIGN
UNIVERSITY OF CALIFONIA BERKELEY | SPRING SUMMER 2012

2045
PUBLIC BOULEVARD INSIDE HORIZONTAL CORRIDOR

NEIGHBORHOOD #1 - TALL COMMERCIAL

NEIGHBORHOOD DESIGN GOALS   CORE PLACEMENT DIAGRAM

2055
REMEDIATION PARK IN EMPTY RAILYARD

首尔的历史包含很多层面，也因此构成了一个丰富多样的城市生态系统。而现在，当经济繁荣发展，老建筑（30至40年历史）被夷为平地来建设新建筑——越来越好、越来越高的新建筑。我们的方案对这种通过拆除进行城市更新是否必要提出了质疑，并提议了一种支持城市不断发展的基础结构。随着居民年龄的增长，城市的年龄也在增加。城市的自我循环将形成一个系统，在这个系统中，人们辛勤地工作、追逐着梦想。

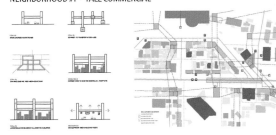

LONGITUDINAL SECTION

The history of Seoul has many layers that constitute a rich and diverse urban ecosystem. Today, the economic boom times have seen old buildings (thirty to forty years old) razed for new developments – always better, always taller. Our project questions that apparent necessity for urban renewal through demolition and proposes an infrastructure that will allow the city to grow incrementally. As the citizens grow older, so does the city. The city's cycles will be seen as a system made up of the peoples' work and dreams.

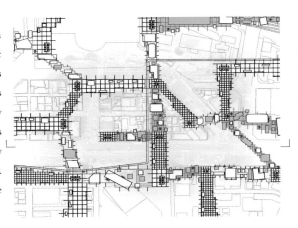

Seamless City
The Chinese University of Hong Kong, Team A

无缝城市
香港中文大学 A组

导师Tutors: HO Puay Peng, Bernard V. LIM JP
设计团队Team members: CHAN Ka Lok,
TSANG Lok Hin Kenneth, TSANG Siu Ha Maggie,
UNG U Tong Annice

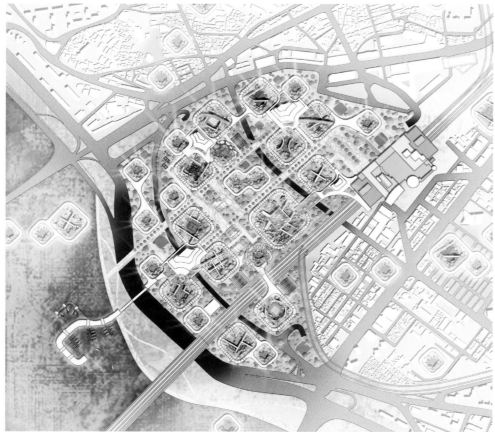

OVERALL STRATEGIC PLAN 2km×2km

在"无缝城市"方案的设计中,首尔市中心为人们提供了连续且相互连接的城市活动空间,促进形成一种全新的多代人的生活方式,将老年人群体也融入社会之中。无缝城市不受时间、速度、地面、以及年龄的限制,其解决方案主要关注于社会问题、经济改革模式和环境发展。基于场地调研和分析,无缝城市提出了一种将韩国传统生活方式和未来居住体系联系起来的新型城市形态。

The centre of Seoul has been designed as a Seamless City that offers a continuous and linkable programmatic urban space. This promotes a brandnew multi-generational lifestyle for a society in which age groups are fused. The Seamless City is not restricted by time, speed, ground, and age. The Seamless City solution focuses mainly on social concerns, economic reform models, and environmental progression. Derived from on-site research and analysis, the solution generates a new urban form connecting the traditional Korean lifestyle to future systems.

Interface City
The Chinese University of Hong Kong, Team B

界面城市
香港中文大学 B组

导师Tutors: HO Puay Peng, Bernard V. LIM JP
设计团队Team members: Albert YUEN, CHEN Chao, Stone CHUNG, Philipp KRAMER

2050　　**2035**　　**2020**

面对低出生率和人口老龄化，我们怎样才能创建一个吸引所有年龄群的有活力的城市呢？界面城市包含三个假设。首先，每个人都是与众不同的，因而他们的需求也有所不同，所以需要一个多样化和充满活力的市场来保障每个人的兴趣——也就是说提倡企业家精神。其次，我们需要重新思考儒家思想，要保持一定数量的临界群体来维持市场多样化，同时促进不同年龄群体间的家庭关系。最后，我们需重新思考城市类型学，使城市与自然共存。

We are faced with a low birth rate and an ageing population. How can we create a vibrant city that attracts all age groups? Interface City incorporates three hypotheses. Firstly, everyone is unique and therefore their needs will be different. Thus, a vibrant and diverse market is needed to guarantee everyone's interest. This could mean that the promotion of entrepreneurship is necessary. Secondly, we need to rethink the idea of Confucianism to retain a certain number of threshold populations that will support a diverse market as well as promote family relationships across age groups. Finally, we need to rethink urban typologies to allow the urban form to coexist with nature.

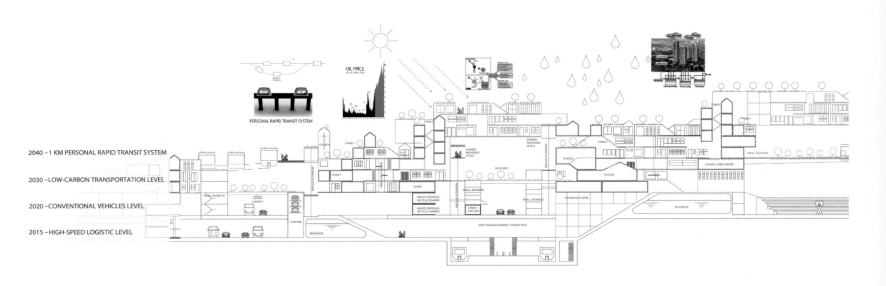

Reviving Urban Spine
Swiss Federal Institute of Technology Zürich, Team B

重振城市主干道
苏黎世联邦理工学院 B组

导师Tutors: Alfredo BRILLEMBOURG,
Hubert KLUMPNER, Kees CHRISTIAANSE,
Michael CONTENTO, Nicolas KRETSCHMANN
设计团队Team member: Youngjin JEONG

我们以对老龄化城市的重新定义作为开头：它是美好生活的催化剂，并有着能经得起时间考验的城市规划。评判一座城市是否适合老龄化，有三个必需的标准：首先是交通便利性，我们规划了一条城市主干道，连接了首尔这座超大城市的主要节点，为居民提供真正的便利交通。其次是包容性：这个主干道也是提供各种不同活动的缓冲区，让周围所有的居民都可以使用它。最后的标准是生态敏感性：一个在社会上富余且生态的城市才是优雅的老龄化城市的代表。

We began with the redefinition of the ageing city; it is the catalyst for even better living quality and at the same time it is a powerful enough urban plan to endure the test of time. We found three criteria especially needed for the aging city. The first is accessibility: we proposed this urban spine, connecting major nodes for this mega citySeoul, provides genuine accessibility for residents. The second criterion is inclusion: the hope is to create a barrier-free zone for myriad diverse and meaningful activities, and people from various surrounding districts can make use of the space. The last one is ecological sensitivity: a city that is socially rich and green at the same time is representative of a gracefully ageing city.

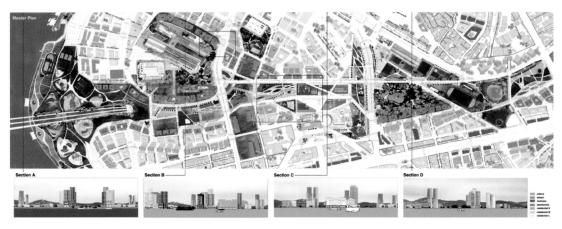

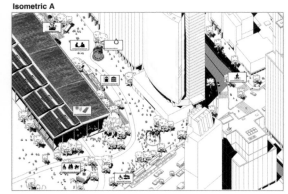

Everyday Urbanism National University of Singapore, Team B

日常的都市生活
新加坡国立大学 B组

导师Tutors:HENG Chye Kiang, WONG Yunn Chii, NG Wai Keen, HWANG Yun Hye, Jurgen ROSEMANN, LOW Boon Liang
设计团队Team members: Sara CHAN, HUANG Jun Cheng, LIM Wei Qi, MONG Wei Meng

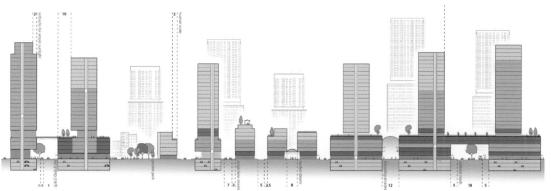

面对老龄化和高密度的人口，日常生活变得更加重要。我们试图保留并弘扬每个人的日常文化，而其关键是要建立个性化并具有相关性的灵活体系。"日常的都市文化"方案在与韩国的街道类型学（在龙山老城区的城市肌理中很常见）相呼应的同时，表明了如何在超高密度环境中与当地现有的交通、绿化、和全球网络潜力相整合，实现一种高品质的生活：生活变得更轻松、更有活力（并不仅仅对老年人而言），从而使城市的每一天都变得非常美好。

In the face of ageing and high density, the realities of everyday life become more pertinent.We seek to retain and promote everyday culture for everyone, and the key is to create flexible systems that can be personalised and kept relevant.Echoing the dynamism of Korean street typologies (the richness of which is prevalent in Yongsan's old urban fabric), Everyday Urbanism shows how a high quality of life can be achieved in a hyper-dense environment, and be wellintegrated with the site's existing strong transport, green, and global network potentials.Living becomes more effortless and dynamic (and not just for the ageing population), making the city good for the everyday.

Permeable City
Tongji University, TeamB

具有渗透性的城市
同济大学 B组

导师Tutors: HUANG Yiru, YAO Dong, ZHU Peidong

设计团队Team members: LI Muzi, TAN Zilong, WU Jing, WU Yunsong, YUE Weilong, ZHAO Jiannan

对于城市的各种弊病，人们普遍认为地表硬化对城市微生态造成的危害最大，其降低了城市地面的渗透性，并会引起洪涝、热岛效应等生态危害。我们的方案意在调解密集的城市化与迫近的生态危害之间的冲突。基于对首尔洪水记录的全面观察和地质条件的研究，我们尝试创造一个以再生水网络为中心的生态系统，以解决首尔恶性城市化带来的多重问题。伴随着水循环周期的再现，我们采用了一个多层表面以安置这个巨型都市——该多层表面是一个密集的基础设施网络，穿梭于整个社会。

Among the urban ills, surface hardening is widely deemed to have the most serious impact on the micro-ecology of the city. Surface hardening reduces the permeability of the urban ground and will cause ecological hazards such as waterlogging, and the heat island effect. Our proposal intends to reconcile the conflicts between intensive urbanisation and the looming ecological hazards. Based on a comprehensive review of the flood records of Seoul and a study on the geological conditions, we attempt to create an ecosystem pivoted around a reclaimed aquatic network to address the multiple issues resulting from ruthless urbanisation in Seoul. Along with the reappearance of the water circulatory cycle, we have adopted a multi-layered surface to accommodate the mega-city, which is a thick infrastructural network that weaves through the whole society.

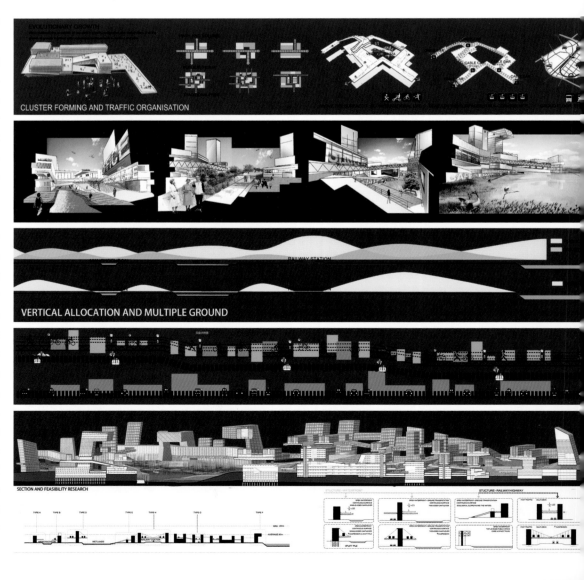

Yongsan Way
University of Michigan, TeamB

龙山的方式
密歇根大学 B组

导师Tutors: Monica PONCE DE LEON，
Lars GRAEBNER
设计团队Team members: Pooja DALAL,
Jonathan MOORE, William TARDY,
Sasha TOPOLNYTSKA, Nathan VAN WYLEN

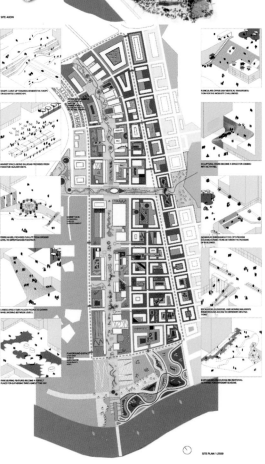

我们通过三个层面的设计来解决现有问题。建筑层面：拒绝根据收入或年龄分群的开发模式，关注于创造多样的住房，而且每栋建筑支持着处于不同生命阶段的人们的生活。街区层面：本项目对建筑插入了垂直的社区空间，并将每栋住宅建筑物置于一个多功能的裙楼上，提供不同层面的服务。区域层面：通过在铁路线上创造架空的系统管道和绿地而充分利用空间。管道系统为社区提供基础设施，同时防止潜在的洪水。其上的连续广场将整个场地交织在一起。

We address these problems at the building, block, and district scales. At the scale of the building, our project rejects the trend toward income or age-specific development and focuses on creating a diversity of housing types within each building that support people in different stages of life.

At the scale of the block, this development requires a number of services for working, living, and entertainment. To address these needs, the project inserts vertical community spaces into the building typologies. We place each residential building on a mixed-usepodium, providing retail and commercial opportunities and larger public spaces for the residents and greater community.

At the scale of the district, wecapitalise on the site's underutilised rail corridor by creating a raisedsystems conduit and public park running over the train right-of-way. The systems conduit provides the district with infrastructure, all while protecting from potential floods. The continuous plaza above weaves the site together.

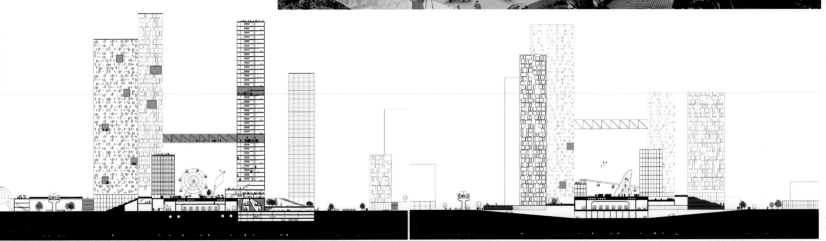

EquaCity
Tsinghua University, Team A

EquaCity
清华大学 A组

导师Tutors: ZHU Wenyi, WANG Hui
设计团队Team members: FU Junsheng, SHEN Feng, WANG Chuan, GUO Leixian

未来亚洲城市的物理形态必须是EquaCity，一座反映平等理念的城市。面对目前首尔市两代隔亲的问题，我们提出将整个都市空间从垂直的EquaCity整合成一个圆圈结构；由此，所有不同年代的人们的都市生活就可以交织在一起。我们的观点是立体城市并非一定要垂直发展；相反，水平格局能更好地适应这种情况。有了衔接的基础设施系统，EquaCity就能顺利地融入场地。

The physical form of future aging society in Asian cities must be an EquaCity, reflecting the philosophy of equality. Seoul is now faced with the problem of intergenerational isolation. In order to solve the problem, we propose the strategy of integrating the whole urban space of the vertical EquaCity into a circle, soall the urban activities of people of different generations can be woven together. As for verticality, our idea is that the vertical city is not bound to vertical growth. Instead, the horizontal urban form, with a tempereddimension of verticality, can adapt to the situation much better. With an articulated infrastructure system, the EquaCity of Yongsan is embedded smoothly into the site.

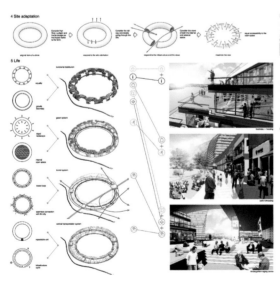
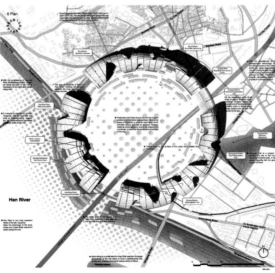
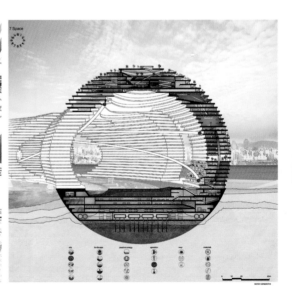

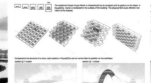

Vertical S(e)oul Living
Tsinghua University, Team B

立体首尔
清华大学 B组

导师Tutors: ZHU Wenyi, WANG Hui
设计团队Team members: ZHANG Qiang, LIU Yun, REN Jie, WAN Bo

首尔目前面临着两代隔亲问题。作为回应，我们创造了首尔（灵魂）立体式生活。鼓励"积极养老"，并加强老年人的独立能力。通过综合设计使得"居家养老"成为可能。我们通过城市格局整合不同年龄的市民，拉近他们的城市生活。同时，我们的设计考虑了如下问题：可持续性、生活质量、技术创新、文脉关系和可行性。这个住宅占总面积50%的项目提供了一种完善的生活、工作、娱乐模式。

Seoul is now facing the problem of intergenerational isolation. In response, wehave createdVertical S(e)oulLivingto de-stigmatiseageing.This design encourages "active ageing" and builds up the ability of the elderly to remain independent by providing a holistic approach to supportive environments. It allows "ageing in place" through inclusive and integrative design. We propose the strategy of integrating all generationsthrough urban form. The urban activities of people of different generations can be drawn together. Our proposal considers issues such as sustainability, quality of life, technical innovation, relationship to context, and feasibility in a holistic and integrated manner. It provides the full slate of live-work-play options, with the residential component making up to 50% of the total floor space.

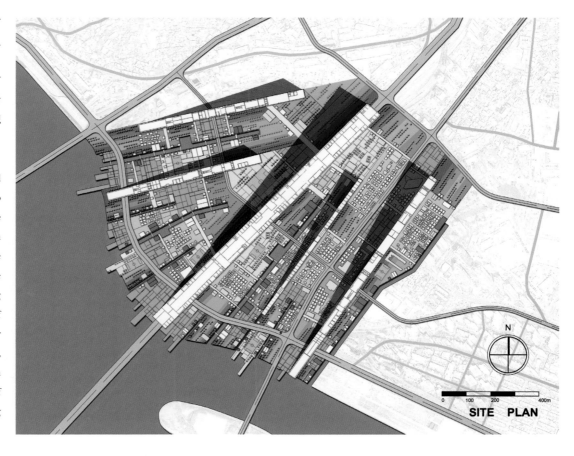

SITE PLAN

SAN
University of Pennsylvania, Team A
山
宾夕法尼亚大学 A组

导师Tutors: Marilyn Jordan TAYLOR,
Matthias HOLLWICH
设计团队Team members: Jordan BARR,
Natali MEDINA, Nicole REAMEY

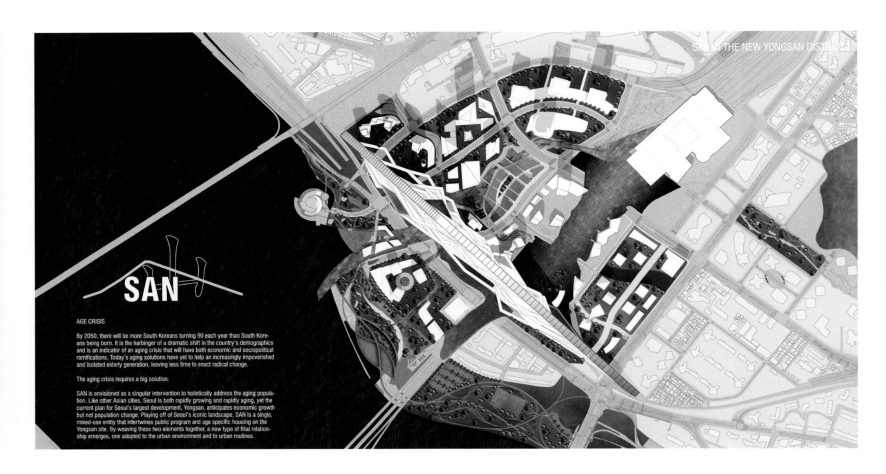

"山"是一个连接了公共设施和不同年龄层住宅的多用途实体,并通过组合这两种元素,形成一种新型的父母和子女关系。

"山"转变了首尔的主要城市形态——复式公寓街区。为了保证居住和流线的高效,新的建筑类型被放大。"山"的商业中心将吸引龙山附近的游客来这里歇脚、享受午餐、或躺卧在绿地,在相互交流间形成一个活跃社区。在集市、影院和博物馆吸引大量客人的同时,"山"的保健服务、运动设备和便利店也能满足居民的日常生活。"山"的理念是:有了综合的服务设施,老年人的生活方式仍可以充满活力。

SAN is a single, mixed-use entity that intertwines public programme and age-specific housing. By weaving these two elements together, a new type of filial relationship emerges. SAN transforms the dominant urban form found in Seoul: the double-loaded apartment block. This new typology – based on efficiency of circulation and habitation – is magnified to great proportions. The commercial core of SAN will attract visitors from around Yongsan to stop for lunch, or lounge on its grassy slopes, kick-starting an active community based on interaction. Markets, theatres, and museums attract visitors to SAN, while healthcare services, exercise facilities, and basic convenience stores supplement the daily routines of the residents. SAN is based on the idea that an elderly lifestyle can still be an active lifestyle, with an integrated backbone of services.

Micropolis
University of Pennsylvania, Team B

微型城市
宾夕法尼亚大学 B组

导师Tutors: Marilyn Jordan TAYLOR,
Matthias HOLLWICH
设计团队Team members: Maru CHUNG,
Nicola MCELROY, Kate RUFE

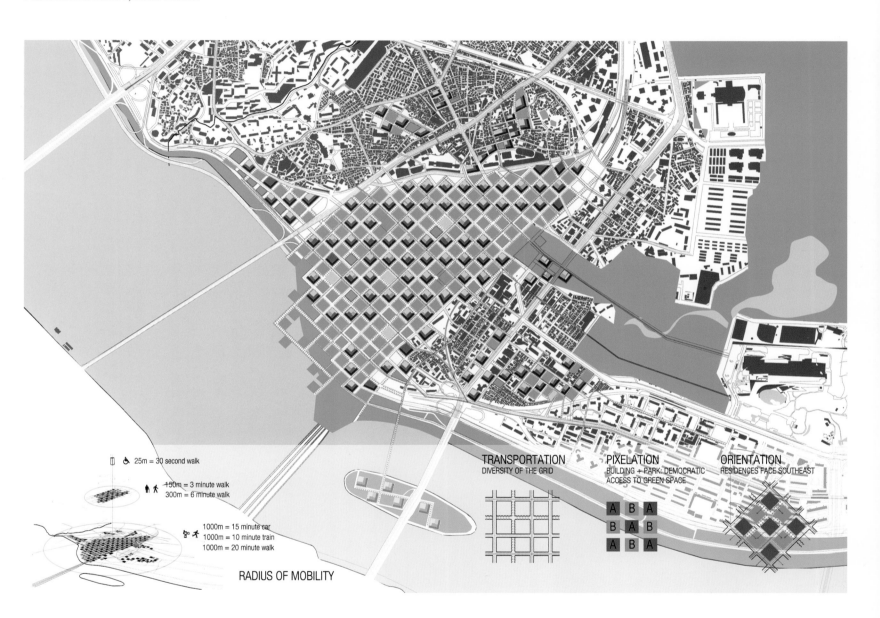

"微型城市"提出一种新的建筑类型——垂直高层建筑类型,以其独立、自然、陪伴和关怀的理念为各年龄层的居民提供全方位的服务。该方案鼓励他们自愿做力所能及的工作,创造一种立足于社区的新型经济。整个场地以像素化的方式形成了棋盘格局,而垂直交通则使所有居民可以方便地享受公园。通过三层不同平面的交通方式,新的城市网络得以进一步强化。在每幢建筑里都配有绿色区域引导空气流通,创建了社区花园以及大型公共空间。

Micropolis proposes a new building typology, vertically layered high-rise typology, that services the full spectrum of needs by bringing the city, with its independence, spontaneity, company, and care, to its multi-aged residents. Micropolis encourages residents to continue working or volunteering to their fullest capacity, creating a new economy based on community. A checkerboard organisation is pixelated throughout the site, creating a vertical density with easy access to park space for all its inhabitants. This new urban grid is further enhanced by a tri-level transportation scheme. Green space is brought inside each building, guiding circulation and creating community gardens as well as large public spaces.

评审团颁奖辞 Jury Citation

我们衷心感谢所有团队所做出的努力以及为支撑他们的方案所做的综合分析。听取大量机智的甚至绝妙的方案，以及出色的演示介绍，对评奖团来说是一项辛苦而艰巨的任务。评审团希望能恭贺所有的团队，但遗憾的是，不可能所有团队都获得奖励。我们将为四个团队颁发优秀奖，并将另外三个奖项颁发给那些具有与众不同价值的方案。

第一个获得优秀奖的作品，我们认为它拥有若干特殊品质。整个方案的规划具有导向性；它展示了城市建筑的一种增量理念。我们欣赏它的众包措施；使老年人能够参与到日常生活中的努力；而且最重要的是，不同催化群集的方案。所以，第一个优秀奖颁给来自苏黎世联邦理工学院A组的方案"老年人催化群"。

第二个优秀奖，我们希望表彰给专注于恢复河道系统，并在一定程度上重建地方的地域身份，同时将城市的其他部分与规划节点相连接的方案。第二个优秀奖颁给来自密西根大学A组的"居住于接缝之中"方案。

第三个要给予优秀奖奖励的，是一个具有非常强大的建筑构思、同时展示了重现历史及创新交通系统的有趣方法的团队。它是个极为有趣的方案。我们为来自东京大学A组的"溪流"方案颁发优秀奖。

第四个优秀奖的颁发，表达了我们对这样一个团队的赏识：试图从城市尺度和社区尺度两个方面来进行协调。他们抓住了生机勃勃的街区生活的特点。第四个优秀奖授予新加坡国立大学A组的"温柔龙山"方案。

现在，将要颁发另外三个奖项。

获得三等奖的项目最值得我们表彰的是，它对规划网格的亲密度和紧凑度的关注。在进行方案设计时，这个方案考虑到了历史的变迁、微尺度敏感整合、旧与新的复杂混合。我们也考虑了执行这个方案如何具有经济可行性。我们将三等奖颁发给来自同济大学A组的"软城市"方案。

用简单、明晰的概念装点城市不失为一项成就，同时，这个团队特别注意到老年人对一个安全、方便的移动系统的需求。在这个方案中，循环圈与城市的其他地方建立了有趣的连接。二等奖颁发给来自东京大学B组的"横向循环"方案。

最后，我们想突出一种有思想的、以过程为导向的方法以及从各方面对城市建筑进行综合考虑的观念，在设计中用独到的眼光看待未来，甚至带有相对开放的形式化。结果和形式固然很重要，但是评审团想强调这些方案为了满足此次比赛要求而对过程和概念的关注。最终，由于两个团队的价值观互为补充，很难判定在这两个全面又周到的方案中，究竟谁更胜一筹。所以评审团决定：来自代尔夫特理工大学的两个团队的"城市无限"和"居住一生的城市"两个方案共同获得一等奖。

We really appreciate the great effort of all the teams, and the comprehensive analysis that supports their proposals. It was hard work for the jury to attend to the incredible amount of intelligent and even brilliant ideas, and to the beautiful presentations. The jury wishes to congratulate all the teams. Unfortunately, it is not possible to award all. We would like to award four teams with commendations, and present three prizes in recognition of some quite different project values.

In the first commendation, we recognise several qualities. This proposal is process oriented; it shows an incremental idea of city building. We appreciate its crowdsourcing; the effort to enable the elderly to participate in daily life; and foremost, the proposal of different catalyst clusters. That's why the first commendation goes to the project "Senior Catalyst Cluster" by Team A from ETH Zürich.

In the second commendation, we wish to praise a proposal that focused on recovering the river system, somehow reconstructing the geographical identity of the place, and also connecting the proposed nodes with other parts of the city. The second commendation goes to "Inhabiting the Seams" by Team A from the University of Michigan.

The third commendation recognises a team that defends quite a strong architectural concept but at the same time shows an interesting way of recovering history as well as an innovative transportation system. It is an Ex'Streamly interesting proposal. We wish to award "Ex' Stream" by Team A from the University of Tokyo.

The fourth commendation shows our appreciation for a team that tried to conciliate two scales – the city scale and the neighbourhood scale. This team captured the flavour of a vibrant street life. Our fourth commendation goes to 'SoftYongsan' by Team A of the National University Singapore.

Now, the three prizes. We wish to award special attention to the intimacy and compactness of the proposed grid of the third-prize-winning project. It enriches the place with consideration of history, a sensitive integration of the micro scale, and a complex mixing of old and new. We also considered how implementation of this proposal would be quite affordable. We award the third prize to "Soft City" by Team A from Tongji University.

It is quite an achievement to encircle a city with a simple and clear concept, and at the same time, to pay mindful attention to the needs of the elderly regarding orientation and a safe and convenient mobility system. In this proposal, there are interesting links between the loop and the rest of the city. The second prize goes to "Trans-Loop" by Team B from the University of Tokyo.

And finally, we wish to distinguish a thoughtful process-oriented approach and a comprehensive consideration of a variety of dimensions that contribute to city building, with a clear vision towards the future – even with a relatively open formalisation. Results and forms really matter, but the jury wants to highlight in these proposals the attention paid to process and programmes in order to face the requirements of the competition. In the end it was very difficult to distinguish between two very comprehensive and thoughtful proposals, due to their complementary values. So the jury decided to propose a shared first prize for "The Open Ended City" and "Lifetime City" by the two teams from Delft University of Technology.

SYMPOSIUM PAPERS
研讨论文

对城市背景下的老龄化问题进行的讨论倾向于鼓励使用两种调查方法。其中一种方法是考量独栋建筑和城市空间如何能够随着人们年龄的增长为人们提供最舒适且最具有包容性的住处；建筑和城市设计如何能够促进"居家养老"来迎合老年人在身体方面的日常需求和鼓励社会包容性呢？另外一种调查方法考虑了城市自身的老龄化情况以及独栋建筑和城市空间随着时间的推移，在不断演变的需求和生活方式的要求之下将如何发生变化；关于城市的居民和城市自身，城市如何能够有意义地步入老龄化？在2012年亚洲垂直城市研讨会上公开的这两种思路穿插了对垂直度和密度进行的更全面的讨论。研讨会论文分属四个宽泛的主题：立体发展方向；人口老龄化空间和建筑老龄化；密度研究；案例分析。这四个主题分类提供了关于亚洲和更远地区的城市发展的有价值的见解。

A discussion of ageing in the context of cities tends to encourage two lines of enquiry. One contemplates how individual buildings and urbanscapes can most comfortably and inclusively accommodate people as they grow older; how can architecture and urban designpromote 'ageing in place', catering physically to the daily needs of the elderly and encouraging social inclusivity? The other line of enquiry considers the ageing of the city itself, and how individual buildings and the urbanscape can change with time and the demands of evolving needs and lifestyles; how can the city age meaningfully with respect to its residents and itself? Both threads surfaced at the Vertical Cities Asia Symposium 2012, interspersed with more general discussions of verticality and density.The symposium papers fall into four broad subject categories that offer valuable insights into urban development in Asia and further afield: aspects of vertical development; spaces for the ageing of people, and the ageing of buildings; the study of density; and project discussions.

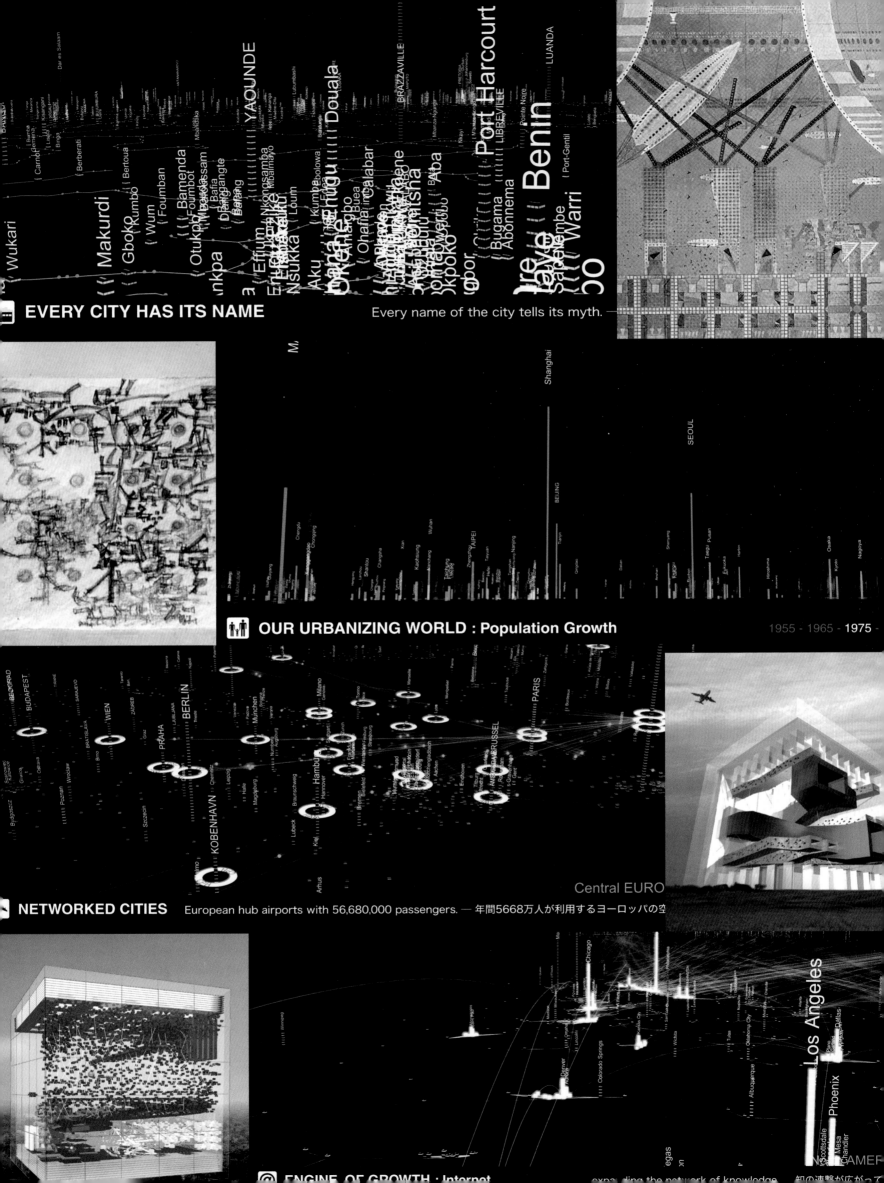

10万人口可持续城市设计标准
Design Criteria for Sustainable Cities of 100,000 People

主题 I：
立体发展方向 1/4
Subject I:
Aspects of vertical development

太田藤原浩
东京大学工业科学研究所 讲师

Hiroshi OTA
Lecturer, Institute of Industrial Science, The University of Tokyo, Japan

自上世纪以来，随着世界城市人口的快速增长，人口超过千万的特大城市的出现常被称为城市化的象征。然而数据显示，超过半数的城市人口居住在人口不超过50万的城市，约10万人口的城市在向大量城市人口提供住房方面起到了很大的作用。通过介绍了一个关于世界城市人口的视频populousSCAPE和名为"500M立方"研究项目，本文强调了10万人口的"小型"城市的重要性及将它们的设计提升至世界城市化前沿的可能性。

With the rapid increase in the world's urban population since last century, the emergence of megacities with over 10 million people is often cited as the symbolic phenomena of urbanisation. However data shows that over 50% of the urban population lives in cities with populations under 500,000, and the cities of around 100,000 people are quantitatively important for housing the urban population. Through the research of a project for the visualization of the world's urban population called "PopulousSCAPE"and a project titled "500M Cube". the paper highlights the importance of "smaller" cities with 100,000 people and the potentials of their design at the frontline of world urbanization.

图1 Fig.1
特大城市在1985(左)及2015(右)
Megacities in 1985 (left) and 2015 (right)

城市人口分布情况

当前，有一半的世界人口居住在城市里。正是人们蜂拥而入的城市造成了对资源的巨大消耗并对周边的自然区域产生了重大的冲击。但正如Wally N' Dow十年前所料，[1]城市不仅仅能将问题集中到一起，还是我们寻求未来的可持续发展的一种有效方法。虽然快速发展的城市化严重影响到了自然环境，但它仍是我们未来共同发展的关键。

然而，我们会面对这样一个问题：城市化过程是怎样进行的呢？人们常说人口超过千万的特大城市正在增加。1985年，有8座人口超过千万的特大城市，在2015年将增加到23座[图1]。发展中国家出现的特大城市（如达卡、马尼拉、德里和卡拉奇）通常被视作世界城市化的最具象征性的现象。然而，目前并无对人口不足千万的城市现状的详细的文档记载，并且这些城市的潜力有时也被忽略了。下文中，我们将通过斯蒂凡·赫尔德斯提供的数据探究城市人口分布情况的更多细节。赫尔德斯创办的网站world-gazetteer.com以通过各国政府人口普查分析报道城市人口数据而著称。这个网站的数据资料提供了45440个城市的人口数字、城市名称、国家名称及经纬度。此处得出的城市人口总数约24亿，在2002年大约占了世界人口的百分之四十[表1]。

但是，如格陵兰最小的城市Qernertuarssuit市（人口只有一个）这个实例，其数据的清晰程度和可靠度有时是值得怀疑的。这是因为赫尔德斯在引用政府人口普查时并未作任何说明——至少在记录城市名称和经纬度时是这样。为了避免不准确性，以下分析只针对人口超过10万的城市[表2].

按面积统计的分布情况

如果我们将数据资料按六个区域进行划分[图2]，将发现亚洲地区城市人口占到世界城市人口总数的一半。在亚洲地区城市人口中，印度是城市化程度最高的国家，城市中心区人口达176 476 551（占亚洲地区城市人口总数的22.8%，世界城市人口总数的11.4%），中国仅次于印度，居第二位，城市中心区人口达163 033 765（亚洲地区城市人口总数的21.1%，世界城市人口总数的10.5%）。如果从中国是世界人口最多的国家这个角度来考虑，那么根据以上数据我们可以得出2002年中国的城市化程度大大低于印度这样一个结论。同样，很有意思的是，拉美地区仅次于亚洲地区。虽然非洲地区排名第四，它仍有可能赶上欧洲地区；据报道，非洲地区城市增长速度最快，达到了3.2%的速度，而欧洲地区的城市增长速度仅为0.1%。

如果我们将各个地区的城市规模进行比较，会发现城市人口的分布情况根据城市化环境的不同而不同。比如，欧洲地区的排名-规模相关曲线较为平滑[图3]，但亚洲地区的大型城市与中型城市之间有着明显的差距[图4]。

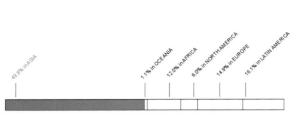
图2 Fig.2　城市人口分布
Urban population distribution

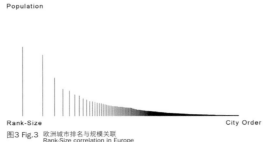
图3 Fig.3　欧洲城市排名与规模关联
Rank-Size correlation in Europe

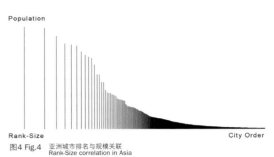
图4 Fig.4　亚洲城市排名与规模关联
Rank-Size correlation in Asia

当我们观察两个案例中城市规模的细分时，就会发现亚洲地区的大型城市比欧洲地区的大型城市更占优势[图5和图6]。

按城市规模统计的分布情况

欧洲地区城市规模的细分说明小规模城市（指人口少于50万的城市）中的城市人口占到了一半。在亚洲地区，大型城市占主要地位。

如果我们对全球人口分布情况分析进行观察，就会发现56.1%的城市人口居住在人口不足百万的城市，40.8%的城市人口不足50万[图7]。如果我们重新观察一下先排除的人口不足10万的城市的人口数据资料，就会发现在城市化过程中，仍然是小型城市在数量上起到了非常重要的作用。

世界银行[2]指出，"发展中国家的城市化仍然集中在小型城市，但大型城市也处于增长势头。"那些在1990年容纳了15%的城市人口的人口超过五百万的城市有望在2015年容纳19%的城市人口。同时，人口不足百万的城市的城市人口比率将从66%下降至59%。

我们可以从这种一级城市和次级城市的集中化趋势推断出两个可能的方向：一是增加这种集中化程度打造能以更加有效的可持续方式使用资源的大型城市；另一个是利用分散政策，促进大量小型城市的增长。

PopulouSCAPE：城市人口可视化项目

根据赫尔德斯提供的数据资料，我们现在可以画出世界城市人口的位置和规模。由于数据资料包含了经纬度坐标，便很容易绘制这45440座城市及其人口特征。

PopulouSCAPE[3]是一个实现这些数据的可视化项目，笔者及其同事们在过去数年[4]间一直在进行这个项目的研究，这个项目主要针对人口超过5万的城市[表3]。

PopulouSCAPE是一个时长10分钟的短片，影片中模拟了一个"城市夜景"，讲述了世界城市人口的分布情况。影片中，十万人口的一座城市作为一个单层建筑单位；百万人口的一座城市作为一个10层的建筑单位；千万人口的特大城市作为一个100层的建筑单位[图8]。

都市夜景从亚洲地区开始，到北美地区结束。伴随着分为四个乐章的原声音乐[图9]。该短片通过人口增长的动态视频、国际航班道路网络及互联网让我们清楚了解在一个"populouscape"中，我们的城市是怎样实现相互联系的。

第一部分讲述了亚洲地区从1955年至2005年间的每隔十年的人口规模——亚洲是全球人口最多且人口增长最快的地区。数据基于联合国预测[图10]。第二部分展示了欧洲地区的重要飞行网络，以说明城市间的跨国联系。主要航空港及对应的出发地和目的地的选择是以国际航班乘客的数量为依据的[图11]。第三部分展示了北美地区的道路网络[图12]。第四部分中，展示了南美、中美及北美地区的特大城市周围的人口密集地带。还对北美地区的主要互联网路进行了展示，以说明城市间的密切的联系[图13]。在短片PopulouSCAPE中，可以很清晰地看到以城市景观的形式展示的各个地区的城市分布情况。亚洲地区是人口最密集的地区，布满了各种规模的城市。城市间的距离也很短——尤其东部、西部和南亚地区最为突出。作为世界上城市最密集的地区之一，东海道特大都市带（从日本东京延伸至大阪，幅员500公里）是其中的一个典型代表。

虽然中东地区城市总数不大，但也有一些大型城市，像德黑兰、巴格达和安卡拉。东欧地区，10万人口或20万人口的小型城市分布比较离散且大多不规则。中欧和西欧地区，各个国家都具有一个独特的人口分布特征图。例如，德国的集聚特征表现为莱茵河-鲁尔河区域；法国人口则高度集中在巴黎；英国的中心城市及这些中心城市的郊区则形成了大都会区。

在非洲区域、南美地区和北美地区，大多数城市都是沿海分布的，而内陆地区分布稀疏。在非洲地区，多数城市的人口仍然不超过一百万，但有许多快速增长的城市，如拉各斯和金沙萨。在南美地区，沿海城市大量集中在圣保罗、里约热内卢和布宜诺斯艾利斯等许多大城市周围。中美地区和北美地区同样有这种带有卫星城的大都市。

传统的主题地图不容易以城市的等级及或全球的角度进行呈现，而制作的主题地图以连续性动画更能详细地表现出了这些特征，但是由于人的大脑在视觉记忆方面比较弱的原因，也不容易获得一个全面的图像。另一方面，虽然最终数据没那么精确，但通过隐喻手段制作的可视化材料以一种更生动更全面化的形式呈现给了我们，这类可视化资料对教育性应用会有所帮助。

500M立方项目：10万人口的居住单元

在PopulouSCAPE中，一座10万人口的城市作为一个建筑单位。将一座10万人口的城市作为一个单位的思维最初来源于一个叫"500M立方项目β版"[5]的项目，作者从2002年就开始同建筑师原广司一起进行这个项目了。这是调查现有的人口约10万的城市的设计框架。综合描述了像土地利用或能源消耗这一类的实际数据，然后使人们能够将这种条件转移到立方体形状的建筑环境中。

该建筑的规划为500×500×500立方米（100层）。规定各层50%的区域为非占用区域。因此，总面积为500m×500m×50层，总计1250公顷。这个"500M立方"的人口密度为每公顷80人，根据标准城市规划，这是合理的。这项研究本身于1990年至1994年期间开展[图14]，旨在研究立方体中空间布局的潜力。从2002年至2004年期间，这项研究又得以重新进行，重新研究的目的是进行数量分析——特别是关于生态问题的数量分析——作为法政大学的一个由原广司和笔者共同领导的半年设计室框架。

实际城市调查

"500M立方"项目以对现有人口接近10万的城市进行调查开始。比如，一个小组对国分寺市进行调查，国分寺市位于东京大都会区西部，人口为109 223。首先，对国分寺市的总体土地利用情况进行调查。在该城市的普查数据中，可以查到该城市总占地面积、建筑面积、农业用地、林地、河流地区、铁路占地、公路占地和公园占地的相关资料[表4]。

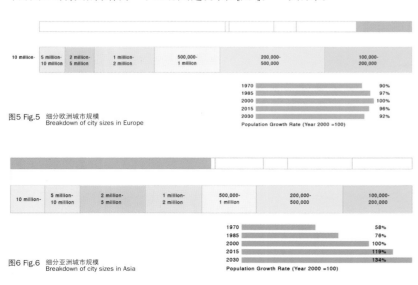
图5 Fig.5　细分欧洲城市规模
Breakdown of city sizes in Europe

图6 Fig.6　细分亚洲城市规模
Breakdown of city sizes in Asia

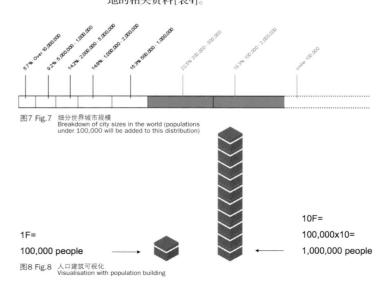
图7 Fig.7　细分世界城市规模
Breakdown of city sizes in the world (populations under 100,000 will be added to this distribution)

图8 Fig.8　人口建筑可视化
Visualisation with population building

但是，关于建筑面积的记录只提供了地面层建筑面积的资料。总建筑面积的计算要通过用地层数量及类目乘上建筑类型，这在详细的城市地图中呈现出来了。

表4显示了"建筑"行列中的转变的区域数据，和相邻城市武藏野市和东久留米市的数据，并且根据"500M立方"的形式使其变得直观，其建筑面积为2.5公顷。例如，国分寺市总建筑容量（8169 300m2）为32.7层[图15和表5]。

在生成立方中的楼层布局的必要数据时，我们同时也对伴随环境影响的问题——如供水、供气、和供电及产生的废弃物——进行了研究。通过这些量化研究，得出了许多惊人的结论。其中之一便是立方的屋顶上的年降雨水量小于各城市的年供水量[图16]。我们可以假定，如果我们循环利用来自立方顶楼的水并通过内部重力作用向底层补充水（比如，作为下水道的污水），这样就可能有足够的水提供给这个"500M立方"。另外一项发现是如果我们将10万人口的城市集中到面积为2.5公顷的区域中，并且将周围的区域空出来用于大型太阳能发电，则将能提供足够的电力来满足城市的需求[图16和图17]。随着光伏板的有效性在过去八年间的大幅提高，需要用于发电的区域面积将比此处计算的小得多。我们同时还对废物量进行了调查。鉴于非易燃废弃物体积相对比较小，地层中将用一两层来装废弃物。

设计输出

在为期四个月的设计工作室的后半期，数组学生将在一个"500M立方"中设计他们自己的10万人口的城市。设计发展过程全程由笔者、原广司教授和环境工程师高井刚明指导，每周举行讨论会。双层框架、内部巨大空白区域的通风系统，重力供水系统和高效运输系统被作为保证"500M立方"的持续性的决定性因素进行讨论。最后得出的研究结果在形式上和理念上有不同的版本，特别是关于住宅单元的规划方法学方面。这种规模的城市有的情况下所含有的城市要素，例如体育场或大学，会成为设计输出的关键设计要素[图18]。

结论

"500M立方"是最终的思维试验。这激发了对城市模式及在最紧凑的区域中城市土地集约利用的讨论。因为该项目是从2002年开始进行到2004年，所以未做足够的环境模拟试验来选取一个最优的模式，但是最近在这一领域的进展可能会给我们一些更加合理的建议。

我们记得，"500M立方"是PopulouSCAPE人口可视化系统中的一个单元，清楚地说明了我们的人口有多稠密，我们消耗的资源量有多大。如本文前半部分讨论过的，城市化一定是我们寻求共同的未来的关键和希望，但是因为这一点，城市的紧密度和效率必将受到质疑和检验。立体发展这个词还有待探索，尤其要从环境的角度进行研究。类似"500M立方"或亚洲立体城能为这类探索提供参考，并充分挖掘建筑师、工程师、研究人员和城市规划者的无限想象力。

图表1 / Table1 — 原始城市人口数据集 / Original urban population data set

Obtained	Cities	Biggest city(population)	Smallest city(population)	Total population
2002	45,440	Mumbai (12,147,107)	Qernertuarssuit (1)	2,412,515,462

图表2 / Table2 — 分析数据集 / Data set for analysis

Obtained	Cities	Biggest city(population)	Smallest city(population)	Total population
2002	3,850	Mumbai (12,147,107)	Targoviste (100,002)	1,547,614,315

图表3 / Table3 — PopulouSCAPE数据集 / Data set for PopulouSCAPE

Obtained	Cities	Biggest city(population)	Smallest city(population)	Total population
2002	8,308	Mumbai (12,147,107)	Valveddittural (50,000)	3,406,869,612

图表4 / Table4 — 东久留米总体用地 / General Land Use in Higashi Kurume (2004)

	Kokubunji (pop.109,223)	%	Musashino (pop. 139,439)	%	Higashikurume (114,706)	%
Architecture	8,169,300m²	65.1%	10,286,598m²	79.1%	7,273,960m²	63.5%
Farmland	2,054,400m²	16.4%	380,873m²	3.0%	2,067,200m²	18.1%
Forest	344,400m²	2.7%	3,480m²	0.0%	193,800m²	1.7%
River	23,000m²	0.2%	96,570m²	0.7%	208,750m²	1.8%
Railway	128,200m²	1.0%	91,097m²	0.7%	43,893m²	0.4%
Road	1,572,800m²	12.5%	1,535,179m²	11.8%	1,440,935m²	12.6%
Park	264,000m²	2.1%	574,064m²	4.4%	219,390m²	1.9%
Total	12,556,600m²	100.0%	13,012,297m²	100.0%	11,447,728m²	100.0%

图表5 / Table5 — 国分寺建筑面积明细 (2004) / Breakdown of area for Architecture in Kokubunji (2004)

	Public	Residential	Commercial	Industrial	Others	Total
m²	236,200m²	747,940m²	114,400m²	332,200m²	7,100m²	1,437,840m²
%	16.4%	52.0%	8.0%	23.1%	0.5%	100.0%

图表6 / Table6 — 国分寺年垃圾量估计 / Volume estimation of annual waste in Kokubunji

	Inflammable	Recyclable	Non-flammable	Bulky	Detrimental	Total
m³	9,860m³	2490m³	1810m³	120m³	17m³	13,897m³
Floors in cube	0.039	0.010	0.006			0.056

图9 Fig.9 PopulouSCAPE: 正在城市化的世界
PopulouSCAPE: night flight over an urbanising world

Distribution of Urban Population

Currently, in 2012, half of the world's population is living in cities. It is the city that people swarm to, creating a huge consumption of resources and heavy impacts on surrounding natural areas. But as Wally N'Dow declared a decade ago, cities are not only concentrating problems, but also the solution for our sustainable future. While rapid urbanisation seriously impacts the natural environment, it is the key for our common future.

However, there is a question that comes to mind: how does this process of urbanisation occur? It is often cited that the number of megacities with populations over ten million is increasing. There were eight such cities in 1985 and there will be 23 in 2015 [Fig. 1]. The emergence of megacities in developing countries (for example, Dhaka, Manila, Delhi, and Karachi) is often mentioned as the most symbolic phenomenon of world urbanisation.

However, the situation of cities with less than ten million people is not well documented, and their potentials are sometimes ignored. In the following section of this chapter, we seek more details of the distribution of urban populations with the data provided by Stefan Helders. The website world-gazetteer.com, maintained by Helders, is well known for its coverage of urban population data from various governmental censuses. Its data set provides population figures, city names and country names, and longitudes and latitudes for 45,440 cities. The total population counted here is about 2.4 billion, which is approximately 40 per cent of the world population in 2002 [Table 1].

However, as seen in the example of the smallest city of Qernertuarssuit in Greenland, which has a population of just one person, the resolution and the reliability of the data are sometimes to be questioned. This is because Helders cites the governmental census without any interpretation – at least to register the name of the city and its longitudes and latitudes. To avoid inaccuracy, the following analyses will be made only for cities with populations over 100,000 [Table 2].

Distribution by Area

If we divide the data set by six regions [Fig. 2], we discover that the Asian region accommodates half of the world's urban populations. Among Asian urban populations, India is the most heavily urbanised country with 176,476,551 people in urban centres (22.8 per cent of Asia's urban populations and 11.4 per cent of the world's), while China is second with 163,033,765 people in urban centres (21.1 per cent of Asia's urban populations, and 10.5 per cent of the world's). When we consider the biggest country in terms of population is China, the data shows that the urbanisation trend in China might have been rather weaker than it was in India in the year 2002. Also, it is interesting to note that Latin

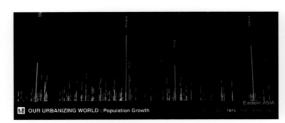

图10 Fig.10 大洋洲及亚洲：人口增长
Oceania and Asia: population growth

图12 Fig.12 非洲：各城均有其名
Africa: every city has its name

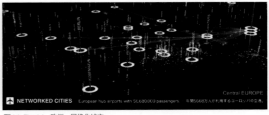

图11 Fig.11 欧洲：网络化城市
Europe: networked cities

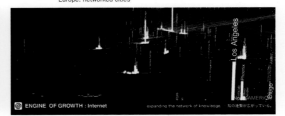

图13 Fig.13 南美及北美：城市集群与交流
South and North America: conurbation and communication

图14a Event 14a

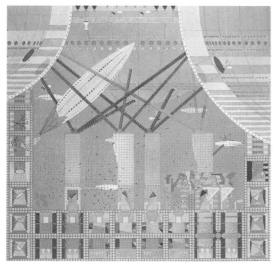

图14b Fig.14b 东京工业大学原广司实验室的 "500M立方" 提案
Proposals for '500M Cube' (1992) by Hiroshi Hara Laboratory, Institute of Industrial Science, University of Tokyo

America follows Asia. Although Africa is fourth, it is likely to overtake Europe; Africa is reported to have the most rapid urban growth rate of 3.2 per cent, while Europe measured 0.1 per cent.

If we compare the city sizes in each area, we find that the distribution of urban populations differs according to the circumstances of urbanisation. For example, while Europe has a smooth curve in the rank-size correlation [Fig. 3], Asia shows a gap between bigger cities and middle cities [Fig. 4]. When we see the breakdown of city sizes in both cases, we can see that the bigger city is more dominant in Asia than in Europe [Fig. 5 and Fig. 6].

Distribution by City Size

The breakdown of European city sizes shows that the smaller-sized cities (those with populations under 500,000) accommodate half of the urban population. In Asia, bigger cities are dominant. If we see the distribution analysis for the whole planet, we can see 56.1 per cent of urban populations are living in cities that have less than a million people, and 40.8 per cent in cities with less than 500,000 people [Fig. 7]. If we recollect the excluded data set of populations less than 100,000, we can see that it is still small cities that are quantitatively playing a very important role in the trend of urbanisation.

The World Bank has noted that "the developing world's urban population is still concentrated in small cities, but large cities' share is increasing." There is the prospect that while cities of more than five million people accommodated 15 per cent of the urban population in 1990, they will house 19 per cent in 2015. Meanwhile, cities with populations of less than one million will decrease from a ratio of 66 per cent of the urban population to 59 per cent.

We can deduce two possible directions from this trend of centralisation to primary and secondary cities. One is to enhance this concentration to form bigger cities with a more sustainable use of resources, and the other is to promote the growth of countless smaller cities with decentralisation policies.

PopulouSCAPE: Visualisation of Urban Population

With the data set provided by Helders, we can now picture the locations and magnitudes of urban populations around the world. As the data set includes longitude and latitude coordinates, it is possible to easily plot the 45,440 cities and their population attributes. PopulouSCAPE is a project to visualise this data, which the author and his colleagues have been developing for some years for cities with more than 50,000 people [Table 3].

PopulouSCAPE is a ten-minute movie showing the distribution of the world's urban population as a 'city nightscape'. A city with 100,000 people is represented as A one-storey building; a city of one million people is represented as a ten-storey building; and a megacity of ten million is shown as a 100-storey skyscraper [Fig. 8]. The night flight over the cityscape starts from Asia and ends at North America, accompanied by original music in four movements [Fig. 9]. Dynamic visualisation of population growth, the network of international flights and roads, and internet linkages let us see how our cities are connected in one 'populouscape'.

The movie was shown in Japan at the Aichi Expo 2005 on an extremely large plasma screen measuring five metres in height and 24 metres in width. After the closing of the expo, screenings were held in Tokyo, Yokohama, London, Linz, Paris, and Daegu to attract international attention.

The first part shows the size of populations every ten years from 1955 to 2005 in Asia – the most populated and fastest-growing region on the planet. Data is based on United Nations projections [Fig. 10]. The major flight network in Europe is depicted in the second part to show transnational connectivity between cities. Major airports and pairs of origin and destination are selected based on the number of passengers on international flights [Fig. 11]. The road network in North Africa is shown in the third part [Fig. 12]. In the forth part, agglomerations around megacities in South, Central and North America are shown. The main internet routes in North America are also depicted to show the vast amount of communication between cities [Fig. 13].

In the movie PopulouSCAPE, each region's distribution of cities can be clearly seen as cityscapes. Asia is the most populated region, crowded with variously sized cities. Distances between cities are short – especially in East Asia and South Asia. One of the world's largest urban concentrations, Tokaido Megalopolis (stretching 500 kilometres from Tokyo to Osaka in Japan), is representative of this.

The Middle East has some large cities such as Tehran, Baghdad, and Ankara, though the total number of cities there is not high. In Eastern Europe, small cities of 100,000 or 200,000 people are distributed discretely and almost uniformly. In Central and Western Europe, there is a characteristic pattern of distribution for each country. For instance, Germany has a characteristic agglomeration known as Rhine-Ruhr; the population of France is highly concentrated in Paris; and central cities and their suburbs form metropolitan areas in the UK.

In Africa, South America, and North America, the majority of cities are located along coasts, while inland areas are sparse. In Africa, most of the cities still have populations of less than one million, but there are rapidly growing cities such as Lagos and Kinshasa. In South America, the coastal cities are heavily concentrated around several megacities such as Sao Paulo, Rio de Janeiro, and Buenos Aires. Central America and North America have also conurbations.

Traditional thematic maps are not easy to depict at city-level or on a global scale. Sequencing animation of thematic maps describes the details better, yet it is not easy to derive a global image, because of the brain's poor visual memory. On the other hand, visualisation using a metaphor gives the detailed data a more lively and more global image, though it is possibly less precise. This kind of visualisation can be helpful for educational use.

500M Cube Project: Dwelling Unit for 100,000 People

In PopulouSCAPE, a city with a population of 100,000

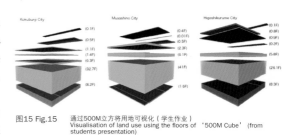

图15 Fig.15 通过500M立方将地可视化（学生作业）
Visualisation of land use using the floors of '500M Cube' (from students presentation)

图16a Fig.16a

图16b Fig.16b 3个城市的年用水总量及2.5公项面积的年降雨量
Annual total water supplies of three cities and the annual rainfalls on 2.5ha area

is represented as a building unit. The idea of a unit for 100,000 people originally came from the project called '500M Cube Project β Version', which the author developed with architect Hiroshi Hara from 2002. It is the design framework to survey existing cities with around 100,000 people. It overviews the actual data like land use or energy consumption, and then allows one to transfer that condition into the cube-shaped building environment.

The building has a plan of 500 metres by 500 metres and a height of the same dimension (100 floors). The floors are defined as having 50 per cent void area. Therefore, the total floor will be 500m x 500m x 50 floors, which means 1,250 hectares. The density of the '500M Cube' is 80 people per hectare, which is reasonable for standard urban planning.

The study itself was initiated in the period of 1990 to 1994 [Fig. 14], and sought the potentials of spatial layout in the cube. It was revisited from 2002 to 2004, the aim being to develop quantitative analysis – especially on ecological issues – as a half-year design studio framework for Hosei University led by Hiroshi Hara and the author.

Survey of Actual Cities

The design studio programme for '500M Cube' began with surveys on existing cities of around 100,000 people. For example, one group surveyed the city of Kokubunji, which is located in the western part of the Tokyo metropolitan area and has a population of 109,223. Firstly, the general land use in Kokubunji was surveyed. In the city's census data, the total area of the city, the built area, and the areas of farmland, forest, river, railway, road, and park can be found [Table 4]. However, the record of the built area only provides the ground floor area of the buildings. The total area for the architecture was calculated through the multiplication of floor numbers and through categorisation by building types, which were given in a detailed city map.

Table 4 shows the transformed area data in the 'architecture' row, along with the data for the neighbouring cities of Musashino and Higashikurume. and visualized according to the format of '500M Cube', which has a floor area of 2.5 hectares. The total architectural volume in Kokubunji (8,169,300m2), for example, is visualised in 32.7 floors [Fig. 15 and Table 5].

While the necessary data for floor layouts in the cube was being produced, issues with environmental ramifications – such as supplies of water, gas, and electricity, and the amount of waste produced – were also researched. From these quantitative studies, several surprising conclusions were obtained. One of them is that the annual volume of rainwater that falls on the roof of the cube is smaller than the annual water supply for each city [Fig. 16]. We can hypothesise that if we recycle water from the top floors of the cube and feed it by the force of gravity to the ground floors (as grey water for sewage, for example), there might be enough volume for the total water supply for the '500M Cube'.

Another finding is that if we concentrate the city of 100,000 people to the area of 2.5 hectares and leave the surrounding areas for mega solar power generation, enough power could be supplied to satisfy the city's demands [Fig. 16 and Fig. 17]. As the efficiency of PV panels has drastically increased in the last eight years, the area needed for power generation may be much smaller than this calculation.

The amounts of waste were also examined. Table 6 shows the huge volume of waste produced by the city of 100,000 people. Considering that the volume of non-flammable waste is comparatively small, the floor accommodating waste would be one or two floors in the basement.

Design Outputs

In the latter half of the four-month design studio, groups of students were to design their own cities of 100,000 people in a '500M Cube'. Design developments were supervised by the author, professor Hiroshi Hara, and the environmental engineer Takeaki Takai through weekly discussions. A double-skinned envelope, a ventilation system for the vast inner void, a gravity-fed water system, and an efficient transportation system were discussed as the decisive elements for the sustainability of '500M Cube'. The final outcomes of the study had variations in forms and ideas, especially on the planning methodology for the residential units. The urban elements that this city size sometimes has, such as a stadium or university, became the key design elements in the outputs [Fig. 18].

Conclusion

'500M Cube' is the ultimate thought experiment. It encourages discussion about the city form and its intensive land uses in the most compact dimensions. As the project was undertaken from 2002 to 2004, the environmental simulations were not examined enough for the seeking of optimal forms, but recent developments in this field may bring more reasoned proposals to us.

As we remember that the '500M Cube' is a unit in the population visualisation system of PopulouSCAPE, it is clearly evident how populous we humans are, and how substantially we consume resources. As discussed in the first half of this paper, urbanisation must be the key and hope for our common future, but for that, the compactness and efficiency of the city must be questioned and examined. The vocabulary for vertical development needs to be explored, especially from environmental aspects. Design frameworks such as '500M Cube' or Vertical Cities Asia may contribute to such explorations, harnessing the unlimited imaginations of architects, engineers, researchers, and city planners.

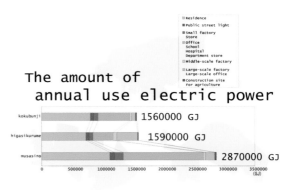

图17a Fig.17a

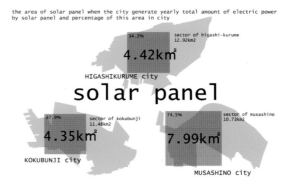

图17b Fig.17b 用电量及对应的光伏板面积（学生作业）
The uses of electricity and interpretation into PV panel areas (from students' presentation)

图18a Fig.18a

图18b Fig.18b 法政大学学生的500M立方提案
'500M Cubes' proposed by the students of Hosei University

参考文献 References
1.Wally N' Dow,W.,联合国人居署中心（HABITAT）中相关"介绍"，一个不断城市化的世界：全球人居报告，1996年，牛津，牛津大学出版社，1996年，21页至24页。WALLY N' Dow, W., 'Introduction' in United Nations Centre for Human Settlements (HABITAT), An Urbanizing World: Global Report on Human Settlements, 1996, Oxford, Oxford University Press, 1996, pp.xxi–xxiv
2.世界银行，转型中的城市：《世界银行城市和区域政府策略》，华盛顿特区，世界银行，2000年。World Bank, Cities in Transition: World Bank Urban and Local Government Strategy, Washington DC, World Bank, 2000
3.福岛，伊藤，川濑，冈部町，及太田，（PopulouSCAPE小组），PopulouSCAPE，漩涡出版社，2003年，第73页至88页。Fukushima, Ito, Kawase, Okabe, and Ota, (Team PopulouSCAPE), PopulouSCAPE, Spiral Publishers, 2005 [DVD]
4.Minami, Y. 和 Ota, H., "PopulouSCAPE: 紧凑城市研究"，10+1，第31卷，日本伊奈出版社，2003年，第73页至88页。Minami, Y. and Ota, H., 'PopulouSCAPE: Compact City Study', 10+1, vol. 31, INAX Publishers, 2003, pp.73–88
5.原广司，Minami, Y. 和 Ota, H., 及日本法政大学学生（2004年），"500M立方项目β版：紧凑城市研究"，10+1，第31卷，日本伊奈出版社，2003年，第89页至108页。Hara, H., Minami, Y., Ota, H., and students of Hosei University (2004), '500M Cube Project β Version: Compact City Study', 10+1, vol. 31, INAX Publishers, 2003, pp.89–108

垂直社区：高密度城市的混合都市形态
Vertical Neighbourhoods: Hybrid Urban Form in High-density Cities

主题 I：
立体发展方向 2/4
Subject I:
Aspects of vertical development

勒内·戴维斯
加利福尼亚大学伯克利分校环境设计学院建筑系教授

René DAVIDS
Professor, Department of Architecture, College of Environmental Design
University of California, Berkeley, U.S.A.

20世纪高度密集的住房规划在很大程度上是标准化、可复制方案——这些方案建立的基础都是旨在平衡个人自由和集体组织的关系的理想化公共计划。通常，这些公共计划都是为不受年龄和疾病困扰的人们服务的。他们与周围大片的空地和公共设施互相隔离，并不存在潜在的个人与集体、公共与私人以及城市与景观之间的辩证逻辑关系——这些规划中绝大多数都只是街区和住房单元的简单集聚而已。本文将对能够与尖端新技术相结合从而建立高度密集却又能适合所有人口居住的社区的住房类型进行普查。重点将是那些既强调建筑和景观的连续性又能界定私人领域且还能建立与更广阔社区相联系的空间的层次结构。

The high-density housing schemes of the twentieth century were largely created from standardised, reproducible plans based on diagrams of ideal communal schemes intended to balance individual freedom and collective organisation. Typically, they were for those unimpaired by age or infirmity. Separated from the surrounding context by swathes of empty land and lacking communal amenities – rather than the promised dialectics of individual and collective, public and private, city and landscape – the vast majority of these schemes were mere aggregations of blocks and units. This paper will survey residential typologies that can be combined with sophisticated new technology to create high-density neighbourhoods suitable for the entire demographic spectrum. Emphasis will be on forms that stress the continuity of building and landscape and the hierarchical network of spaces that defines the private realm and establishes links to the wider community.

亚洲新兴经济的快速发展导致了众多高层住宅群的急速建设，以便容纳大量涌入城市的移居者，而其中一些建筑与储藏库相差无几。这样的密度可能是经济发展不可避免的结果，但是许多人也把它看作是创造创新、节能和宜居城市环境的关键[图1][1]。

勒·柯布西耶倡导将高密度城市作为最大程度利用光线、新鲜空气和开放空间的集中建筑形式。简·雅各布斯在基于性能的分析中表达其对传统城市构造的欣赏，这种分析根据主要由行人活动定义的标准评估邻里环境，但这些争论并非一定彼此不容。[2]本文评估了市区扩展对于在快速发展的东亚城市创造成功居住环境的三个重要方面：交通网络、开放空间和社区网络。[3]

交通网络

城市的垂直发展激起许多建筑师和规划师对空中集合城市的设想，其中空中的空间与大城市的陆地空间具有同等的性质和种类。尽管这些愿景让人着迷，但却仍然难以实现。二十世纪中期，连接高密度住宅的城市高架人行道网络普及全球。其目的是复制更高水平的街道生活，同时将人行道与汽车交通隔离。在现代住宅群落的"空中街道"未能成功区别出公共和私人空间后，这种理念还被看作是社会问题猖獗的导火索。

在城市规模上，城市高架人行道网络被称为人行天桥或高架公路，与居民区"空中街道"类似。明尼阿波里斯于1962年正式推出首个高架公路体系，这一理念随后被美国和加拿大超过三十座城市采用，特别是寒冷气候地区。总的说来，结果有好有坏，因为将人行道和地面活动远离大街后丧失了活力，许多美国市区更是平添几分空旷感。

香港集中使用的中区行人天桥系统取得成功，因为香港的人口密度较高，使用高架路、自动扶梯和桥梁能够缓解地面街道的拥堵现象，而不是减少本来就不多的行人流量，而这正是明尼阿波里斯和其它北美洲城

图1 Fig.1 亚洲垂直城市：重庆
Asian Vertical City: View of Chongqing, China

图2 Fig.2 吉隆坡双子塔（西萨·佩里）连桥内部
Interior of bridge, Petronas Towers, Malaysia by Cesar Pelli

图3 Fig.3 世贸中心竞赛设计（联合建筑）
United Architects' World Trade Center Competition

图4 Fig.4 卡塔尔多哈阿尔法项目（Xavier Vilalta工作室）
Alpha Project, Doha Qatar, by Xavier Vilalta StudioCompetition

市所遭遇的情况。首尔一个成功的小型体系Gwando/Sewoon Arcade由建筑师Swoo Geun Kim设计，方案采用城市多功能混合用地，而不是像欧洲、北美洲和南美洲城市通常使用单一用途住宅或商业分区，这使得这一设计更加成功。④

在世纪之交，新的摩天大厦在高层建筑与桥梁和人行道连接重新唤起了二十世纪中期"空中街道"的愿景。最早的是吉隆坡由西萨·佩里（1994年）设计的双子塔[图2]，两栋楼之间由一座桥连接，可供出入并用作备用逃生通道。后来出现更大胆的设计，比如联合建筑师联盟参加的世贸中心竞赛（2002年）[图3]和斯蒂文·霍尔设计的东直门当代MOMA（2009年），用八座天桥连接八座塔楼。OMA和奥雅纳工程顾问公司设计的中央电视台总部大楼（2012年）将水平和垂直元素融入一个空间和结构连续体。新加坡滨海湾金沙酒店的空中花园（2011年）用一条悬臂式高架路连接三座塔楼，包括一个公共眺望台、花园、饭店、游泳池、慢跑步道，并且可以欣赏新加坡港全景。⑤其它近期项目则更进一步验证了这一趋势，比如比亚克英格斯集团的以首尔标签为形式的交叉#字塔楼和巴塞罗那建筑师泽维尔维拉塔工作室竞赛设计的多哈卡塔尔塔楼（2012年）[图4]。

随着超高层建筑和人口密度的增加，建筑之间的连接变得越来越重要，建筑应能够实现长距离快速垂直和水平交通。随着在单个建筑内部进行的日常活动越来越多，同时旅行距离缩小，三维公共交通网络可以为日益增加的老年人和特殊需求群体提供与安全、更佳气候控制和娱乐选择有关的更高生活质量优势。但是建造这些网络，城市必须实施新的指南，以整合成为较大的网络，而不是零碎的松散个体。⑥

现有的地面交通体系已不能满足高密度城市的要求。扩展的三维网络能够纳入更多的地铁和更快的地下公路连接，比如Costanera del Norte[图5]，一条私人融资的新收费公路，从智利圣地亚哥北部穿过，连接城市东部和西部。该地下公路系统的可行性随着更新、更有效的钻孔技术的发展逐渐增加，自动化也大大提高其安全性、速度和在大型高密度城市运输人员及货物的能力。与地下停车场结合时，该系统有潜力增加大都市交通能力并大大削减城市通勤时间、地面交通走廊专用空间、街道公共运输车数量以及空气和噪音污染。

目前专门用于交通网络的部分表面区域可以重新用作公共空间和种植用途。⑦

绿色网络

绿色空间对于营造城市的特色和建筑肌理以及交通模式都很重要，人们对于在密集城市环境中接近自然众多优势的认识也越来越强。绿色空间可以调节环境的温度和湿度，并且在水循环、清洁污染物和支持各种各样植物和动物的生态系统中起着重要的作用。从生态学角度看，绿色空间是城市唯一的生产性区域。除街道树木、绿化安全岛、公园和地区自然保护区之外，以露台、绿化外立面和屋顶的形式提供的建筑绿色空间在当代住宅设计中成为经常出现的主题，并因人们对环境质量的日益关注而获得重视。⑧

勒·柯布西耶通过设计Immeubles Villas（1922年）引入带有集成型绿色空间的公寓建筑理念，以各单元宽敞的两层高的露台为特征[图6]。莫瑟·萨夫迪的Habitat（1967年）创新性地利用了该理念，建造了一个由带花园的模块化住宅单元叠加而成综合结构。自此数十年来建立了该配置的许多版本，比如戈达德·曼顿在东伦郭Barrier Point设计的豪华住宅楼群，跟Habitat一样面朝海滨。最近，类似的理念应用于比勒·柯布西耶和萨夫迪的设计更高、更密实的结构。针对新加坡的纽顿轩公寓（2007年），WOHA建筑事务所的负责人Wong Mun Summ和Richard Hassell在每幢建筑的第四层都设计了很大的室外公用区。他们在每一楼层都创造了大阳台和较小私人阳台的重复模式。米兰目前在建的住宅楼，由Stefano Boeri Architetti设计的Il Piccolo Bosco Verticale [图7]在每个公寓都设计有私人阳台，呈现为城市绿化植被的垂直轴。

这些项目中最为创新的就是OMA的奥雷·舍人设计的Interlace Complex（在建），背离了新加坡的标准独立、垂直公寓楼类型，而是探寻广阔的、互联的半私人和共享空间网络，结合了私人与公共屋顶花园。绿色屋顶可以缓解城市热岛效应，人口密集地区的绿色走廊也能降低建筑的总体能量消耗，这一设计为高楼大厦密集区的环境和社会问题提供了良好的解决方案。

杨经文[图8]与卢埃林·戴维斯·耶安格建筑事务所和TR Hamzah & Yeang进行的相关探索集中于垂直绿色都市生活的高楼类型学的潜力，他们将带有半公共空间的高

楼内部打开，将植物坡道和人行道重释为花园露台不时点缀的线型公园。勒·柯布西耶的作品以技术和自然明显分界为特色，如果这个特色在杨经文的项目中减弱，那么它在MVRDV的Gwanggyo Power Center City（2008年）中几乎已经完全消失[图9]。它冒着失去城市建筑完整性的风险使自然和人工的差别模糊化。

所有这些项目都创造了新的高水平绿色网络。但是，就所有的创新而言，这些巨大的雕塑物不能提供真实的城市连接。在创新的同时，许多新的设计只是针对享有特权的小型社会部门的特殊解决方案，并不是经济适用的住宅计划，更不能为城市提出综合绿色愿景，从而加固其作为标志性建筑的地位而不是城市未来综合绿色战略的一部分。比如纽约高线公园或巴黎绿荫步道等项目都开发于未使用的高架铁路线，它们带来的只是将新设计融入现有绿色空间网络的启发，以形成与半公共和公共绿色空间以及交通网络三维体系的连接。

社区网络

随着城市继续垂直增长、私人建筑在城市容积区内占着越来越大的比例，其中的半公共和公共空间也构成了较大比例的城市公共空间。大型住宅楼的半公共空间极少能拥有传统广场、城市广场、公园、市场、公有领域、购物中心、村庄广场甚至是机构建筑内部的公共空间（如机场等）的活力，这对于它们而言是一个重大的挑战。现代住宅社区公共空间最早最知名的项目由勒·柯布西耶设计，他赞美了卡尔图哈修道院内公共和私人空间的平衡并且在马赛Unité d'Habitation[图10]（1952年）对它进行了重新诠释。这栋十二层的公寓楼群包含一所学校、儿童设施、娱乐空间和运动设施——跑道、游泳池和位于屋顶的体育馆，在中层还有一条室内商业街。⑨ Unité地处景色壮观但相对隔离的位置，通过底层架空柱抬升，其社会和商业设施从一开始就毫无生机，因为既不能吸引到偶然路过的行人而自身居民又太少不能确保商业成功。

六十年后，MVRDV的马德里之外的Mirador项目（2005年）[图11和图12]在地面以上四十米的半公共广场表达了一种强有力的姿态，一改多层重复单元的单调性。但是由于居民不愿意让孩子到空中十二层的柏油路面玩耍，该空间的功能并不比勒·柯布西耶Unité的屋顶要好。MVRDV在马德里的另一个住房项目Celosia（2009

年）[图13]有许多由三个单元共享的特色小聚集空间。由于不属于任何一个单元，居民并不会使用而且不喜欢这些空间。⑩

东京的Shinonome Canal Court（2003年）由著名建筑师山本里显、伊东丰雄和隈研吾设计，除了共享的公共空间之外，该项目极力避免孤立于地面以及类似勒·柯布西耶Unité和MVRDV马德里项目存在的半公共/私人空间问题。Shinonome Canal Court利用了更加传统的设计，第一层和第二层都有公共空间和露台，将个人单元连接到通往超市、河滨公园、商店、服务、幼儿园和游乐场的蜿蜒主街。该项目足够灵活，能容纳小办公室和公共区域，其隐私度可以用可移动木板调整。⑪

Sargfabrik，由棺材厂改建成的住宅（1996年）和Miss Sargfabrik（2000年），两者都位于维也纳，由BKK建筑师设计，以流行而成功的共享空间为特色，包括文化设施、餐厅、幼儿园、二十四小时浴室（也向公众开放）、社区厨房、排练厅、图书馆和会议及研讨会会议室。这些空间都可以用于举行社区活动，残疾人员也能进出。Sargfabrik和Miss Sargfabrik公共空间的设计并不一定能够照搬到其它项目上，因为其开发人员本来打算创建一个社区并为此成立了一个协会。但是，很明显，这两个项目成功的一个原因是它们与较大社区的类似空间连接较近，而从城市活动中独立出来的公共空间则不太容易与人们的日常生活融合。

20世纪50年代在纽约出现一种与私人办公室和商业建筑公共空间有关的有趣现象。斯基德莫尔、奥因斯和梅里尔将利华大厦（1952年）[图14]的广场和大厅被设计为公共艺术空间，而没有重复典型的顶楼退后，也没有根据当地区划法建造公共庭院。密斯·凡·德罗设计的西格拉姆大厦（1958年）穿过派克大街，以一个大型花岗岩公共广场为特色，同时也符合纽约市1916年区划决议的建筑退后要求。

1961年，该市颁布了区划法的主要修订版，向开发商提供激励政策，让他们修建类似于利华大厦和Seagram建筑项目的公共空间。所谓的"奖励法"鼓励私人开发商提供公共空间，这样城市就会批准比原先允许的面积更大的建筑面积。利用这一条款的项目案例之一是菲利浦·约翰逊和约翰·伯奇设计的索尼大厦（1984年），当时该大厦为美国电话电报公司所有，其中有一个覆盖的拱廊和三层楼的通信博物馆。另一个例子是麦迪逊大道590号（1983年），前身为IBM大厦，由爱德华·拉拉比·巴尔内斯联合公司设计，其公共室内竹园就像一个大大的温室，占了绝大部分建筑面积。由诺曼·福斯特设计的香港汇丰银行总部大厦（1983年）使用了类似的规划策略，第一层以循环走廊的形式通向外界。每到周末——特别是星期天香港各处的菲律宾女佣聚集在那里，坐在地板上开始社交的时候，它会变得格外的活跃。

结论

公共和半公共空间对于儿童、青少年、移民和老年人尤其重要，但是空间的成功取决于设计的质量。当空间与现有的三维交通和人行道网络相交时，更有可能取得成功。最近与先进新技术和材料有关的建筑创新提供了试验的沃土，只要结果不是独立试验即可，它们应该有利于城市高架、地面和地下综合三维连续体。

The rapid growth of emerging economies in Asia has resulted in numerous clusters of hastily constructed residential towers, some of which are little more than storage crates, to accommodate a huge influx of migrants to Asian cities. The resulting densities may be an inevitable consequence of economic development but are also perceived by many as key to the creation of innovative, energy-efficient, and liveable urban environments [Fig. 1]①

Le Corbusier advocated high-density cities as a way of concentrating built form to maximise access to light, fresh air, and open space for urban populations. Jane Jacobs' admiration for the fabric of the traditional city was expressed in a performance-based analysis that evaluated neighbourhoods according to criteria defined mainly by pedestrian activity,② but these polemics are not necessarily at odds with each other. This paper assesses three aspects of urban growth fundamental to the creation of successful living environments in the rapidly growing cities of east Asia: transportation networks, open space, and community networks.③

Transportation Networks

Vertical growth in cities has prompted many architects and planners to imagine aerial conurbations with spaces of equivalent character and variety to those found on the ground in great cities. While these visions are compelling, their realisation remains elusive. In the middle of the twentieth century, networks of elevated urban walkways connecting high-density residential buildings were popular worldwide. The aim was to replicate street life at higher levels while separating pedestrians from automobile traffic. The concept lost favour after "streets in the air" built at modernist housing estates failed to make distinctions between public and private space and became identified as a cause of their rampant social problems.④

At city scale, the elevated urban pedestrian networks known as skywalks or skyways were similar to the residential 'streets in the air'. Minneapolis launched the first skyway system in 1962 and the concept was later adopted by over thirty cities in the United States and Canada, particularly in cold-climate areas. On the whole the results were mixed, as removing pedestrians and ground-level activities from the streets also deprives them of vitality and adds to the sense of emptiness prevalent in many American downtowns. Hong Kong's intensively used Central Elevated Walkway is successful because the city's much higher population density allows the network of skyways, escalators, and bridges to relieve congestion on the surface streets, rather than further reduce already meagre pedestrian traffic, as is the case in Minneapolis and other North American cities. The success of a smaller system in Seoul, the Gwando or Sewoon Arcade designed by the architect Swoo Geun Kim, was enhanced by the decision to implement mixed-use zoning for the surrounding urban fabric rather single-use residential or commercial zoning as was typically the case in European, North American, and South American cities.

At the turn of the millennium, new high-rise towers connected at their upper levels with

图6 Fig.6 别墅大厦（勒·柯布西耶1922）
Le Corbusier, Immeuble Villas (1922)

图5 Fig.5 圣地亚哥哥斯塔内拉地下快速路
Underground Highway La Costanera in Santiago

图7 Fig.7 博斯科立体森林（Stefano Boeri）
Piccolo BoscoVerticale by Stefano Boeri

图8 Fig.8 新加坡热带地区生态设计大厦（T. R.哈姆扎 & Yeang）
EDITTTower, Singapore by TRHamzah&Yeang

图9 Fig.9 广桥能量中心（MVRDV 2008）
MVRDV'sGwanggyo Power Center of 2008

图10 Fig.10 马赛公寓商业街（勒柯布西耶）
Le Corbusier, commercial street, Unitéd' Habitation, Marseilles

bridges and walkways recalled mid-twentieth-century visions of 'streets in the air'. The first of these, the twin Petronas Towers (1994) in Kuala Lumpur designed by Cesar Pelli [Fig. 2], are linked by a single bridge that provides access and alternative escape routes between the two structures. More daring designs followed, such as the United Architects' entry to the World Trade Centre competition (2002) [Fig. 3] and Steven Holl's Linked Hybrid Building (2009) in Beijing, which joins eight towers with eight sky bridges. The CCTV Headquarters Building (2012) in Beijing designed by OMA and Arup melds horizontal and vertical elements into one spatial and structural continuum. The SkyPark at Marina Bay Sands (2011) connects three towers with a cantilevered skyway, which includes a public observatory, gardens, restaurants, swimming pools, jogging paths, and panoramic views of Singapore Harbour.[5] Other recent projects, such as Bjarke Ingels Group's Cross# Towers proposal in the form of a hashtag for Seoul, and Barcelona architects Xavier Villalta Studio's competition design for a tower for Doha Qatar (2012) [Fig. 4], provide more evidence of the trend.

Connections between buildings become more important as the number of super-tall high rises and population densities increase and buildings are required to enable rapid vertical as well as horizontal movement over longer distances. As more daily activities take place inside individual buildings and travel distances decrease, three-dimensional public access networks could offer quality of life advantages related to safety, climate control, and recreational options for the growing numbers of elderly people and those with special needs. However, to build them, cities will have to implement new sets of guidelines so that larger networks can be assembled rather than piecemeal assortments of fragments.[6]

Existing ground-level transportation systems are not capable of meeting the requirements of high-density cities. Expanded three-dimensional networks could include more underground trains and faster underground highway connections, such as the Costanera del Norte [Fig. 5], a new privately financed toll-way passing underneath the northern section of Santiago, Chile, and connecting the eastern and western parts of the city. The feasibility of such underground highway systems has increased with the development of newer, more efficient boring techniques, and automation has significantly increased their safety, speed, and capacity for transport of people and cargo in large high-density cities. When combined with underground parking structures, such systems have the potential to increase metropolitan traffic capacity and significantly reduce urban commute times, the space dedicated to surface transportation corridors, the presence of public transportation vehicles on the streets, and air and noise pollution. Portions of surface areas currently dedicated to transportation networks could be repurposed for public space and vegetation.[7]

Green Networks

Green spaces are as fundamental to the identity of individual cities as built fabric and modes of transportation, and there is growing awareness of the many advantages of proximity to the natural world in dense urban settings. Green spaces condition the environment by moderating temperature and humidity, and play a significant role in the water cycle, cleansing pollutants and supporting a variety of plants and animals. From an ecological perspective, green spaces are the only productive areas of a city. Beyond street trees, vegetated medians, parks, and regional nature reserves, the provision of green spaces in buildings in the form of terraces, as well as vegetated facades and roofs, has become a recurrent theme in contemporary residential design, gaining momentum because of rising concerns about environmental quality.[8]

With his design for the Immeubles Villas (1922), Le Corbusier introduced the concept of an apartment building with integrated green spaces, featuring ample double-height terraces for each unit [Fig. 6]. A substantially modified version of the concept was adopted by Moshe Safdie's Habitat (1967) – a complex of stacked modular housing units with attached gardens. Many versions of this configuration have been built in the decades since, such as the blocks of luxury housing at Barrier Point in East London designed by Goddard Manton, which, like Habitat, also faces the waterfront. Recently, similar concepts have been adapted to structures much taller and denser than those designed by Le Corbusier and Safdie. For the Newton Suites in Singapore (2007), principals of the architectural firm WOHA, Wong Mun Summ and Richard Hassell, designed large outdoor common areas on every fourth story. They created a repeating pattern of large balconies and smaller private balconies on each floor. Il Piccolo Bosco Verticale by Stefano Boeri Architetti [Fig. 7], a residential complex currently under construction in Milan, has private balconies for each apartment and presents itself as a vertical shaft of vegetation in the city.

One of the most innovative of these projects is the Interlace Complex (under construction) designed by Ole Scheeren of OMA, which departs from Singapore's standard typology of isolated, vertical apartment towers and explores instead an expansive, interconnected network of semi-private and communal spaces integrated with vegetation and private and public roof terraces. As green roofs can mitigate the urban heat island effect and green corridors in densely populated areas have been shown to reduce overall energy consumption in buildings, this design provides a good solution to both the environmental and social problems of high-rise buildings in dense areas.

Related explorations undertaken by Ken Yeang [Fig. 8] with architectural firms Llewelyn Davies Yeang and TR Hamzah and Yeang focussed on the potential of the high-rise typology for vertical green urbanism by opening the interiors of tall buildings with semi-public spaces, reinterpreting vegetated ramps and pedestrian walkways as linear park-scapes punctuated by garden terraces. If the clear demarcation between technology and nature that characterised Le Corbusier's work has

图11 Fig.11 马德里马拉杜项目(MVRDV)
MVRDV's Mirador project, Madrid

been softened in Ken Yeang's projects, it has been all but eliminated at MVRDV's Gwanggyo Power Center City (2008) [Fig. 9], which resembles a collection of hills. By blurring the distinction between the natural and artificial, it risks losing its integrity as urban architecture.

All of these projects create new high-level green networks. However, for all their innovations, they are giant sculptural objects that fall short of providing genuine urban connectivity. While inventive, many of the new designs represent exceptional solutions for a small privileged sector of society. None are affordable housing schemes, and they all fail to put forward a comprehensive green vision for the city, thus reinforcing their status as isolated formal icons rather than parts of an integrated urban green strategy for the future. Projects such as the High Line in New York or the Promenade Plantée in Paris, both promenades developed on de-commissioned elevated railway lines, suggest ways in which new designs can be integrated with existing networks of green spaces to form connections with three-dimensional systems of semi-public and public green spaces and transportation networks.

Community Networks

As cities continue to grow vertically and private buildings constitute an increasingly larger proportion of the urban volumetric area, semi-public and public spaces within them will also comprise a larger proportion of urban public space. The liveliness of traditional plazas, town squares, parks, marketplaces, public commons, malls, village greens, and even public spaces within institutional buildings such as airports has rarely been achieved in the semi-public spaces of large residential buildings, and they represent a significant challenge. The earliest and best known examples of projects with public spaces for residential communities in the modern era were designed by Le Corbusier, who admired the balance of collective and private space in Carthusian monasteries and reinterpreted it in the Unité d'Habitation [Fig. 10] in Marseilles (1952). This twelve-storey apartment block contains a school, childcare facilities, recreational space, and sporting facilities – a running track, pool, and gymnasium – located on the rooftop, and an interior commercial street with shops at mid-level.[9] Located on a spectacular but isolated site and lifted off the ground on pilotis, the Unité's social and commercial facilities languished from the start because they were unable to attract any casual passers-by and there were too few residents to ensure commercial success.

Sixty years later, the semi-public plaza forty metres above the ground in MVRDV's Mirador Project (2005) outside Madrid [Fig. 11 and Fig. 12] created a strong formal gesture to relieve the monotony of stacked repetitive units. However, the space has failed to function any better than Le Corbusier's Unité rooftop because residents are reluctant to allow their children play on a tarmac twelve stories up in the air. Celosia (2009) [Fig. 13], another Madrid housing project by MVRDV, features many small gathering spaces shared by as few as three units. As these spaces do not belong to any one of the units, they are not used and are disliked by the residents.[10]

Shinonome Canal Court (2003) in Tokyo, with blocks designed by renowned architects Riken Yamamoto, Toyo Ito, and Kengo Kuma among others, includes communal spaces but avoids the isolation from the ground and the semi-public/private space issues present in Le Corbusier's Unité and the MVDVR Madrid project. Shinonome Canal Court utilises a more traditional design with public areas and terraces located on both the first and second levels, which link the individual units to a winding main street connected to a supermarket, a riverside park, shops, services, a kindergarten, and playgrounds. The project is flexible enough to accommodate small offices and common areas whose level of privacy can be adjusted with movable wood panels.[11]

The Sargfabrik, a residential conversion of a coffin factory (1996) and Miss Sargfabrik (2000), both in Vienna and designed by BKK Architects, feature popular, successful communal spaces including a cultural facility, restaurant, kindergarten, a twenty-four-hour bathhouse (which is also open to the public), a community kitchen, rehearsal room, library, and a seminar room for meetings and workshops. All these spaces are available for the hosting of community events and are accessible by those with disabilities. The design of the common areas of the Sargfabrik and Miss Sargfabrik are not necessarily transferable to other projects because they were developed by a group of people who intended to create a community and formed an association dedicated to that purpose. It is clear, however, that one of the reasons for their success is their close connections with similar spaces in the larger community, whereas public places separated from urban activities are not easily integrated with people's daily lives.

An interesting phenomenon related to public space in private office and commercial buildings emerged in New York in the 1950s. Rather than repeat the typical stepping back of the top floors and creation of a public courtyard to comply with local zoning laws, Skidmore, Owings & Merrill designated the plaza and lobby of Lever House (1952) [Fig. 14] as public art space. Several years later, Mies van der Rohe's Seagram Building (1958) just across Park Avenue featured a large granite public plaza, also to bring the buildings into compliance with the setback requirements of New York City's 1916 Zoning Resolution.

In 1961, when the city enacted a major revision of the zoning law, developers were offered incentives to install public spaces comparable to those of Lever House and the Seagram Building in their projects. The so-called 'Bonus Law' encouraged private developers to provide public space in exchange for which the city would allow larger amounts of square footage to be built on a given site than would otherwise be permitted. An example of the projects that took advantage of this provision is the Sony Building (1984) designed by Philip Johnson and John Burgee, owned at the time by AT&T, which provided a covered arcade and three-story communications museum. Another example is 590 Madison Ave (1983), formerly the IBM Building, designed by Edward Larrabee Barnes & Associates, whose public interior bamboo garden resembles a large greenhouse and occupies most of the building's footprint. Using a similar planning concession, the first floor of the Hong Kong Shanghai Banking Corporation's Headquarters Building (1983) designed by Norman Foster is available to the public as a circulation corridor. It becomes intensely active on the weekends – especially on Sundays, when Filipino housemaids from all over Hong Kong congregate there and sit on the floor to socialise.

Conclusion

Public and semi-public spaces are particularly important for children, adolescents, immigrants, and the elderly but the success of these spaces depends on the quality of their design. They are more likely to be successful when they intersect with existing three-dimensional transportation and pedestrian networks. Recent architectural innovations combined with advanced new technologies and materials provide fertile grounds for experimentation, as long as the results do not exist as isolated experiments. Rather, they should contribute to a comprehensive three-dimensional urban continuum of elevated, ground-level, and subterranean spaces.

图12 Fig.12 马德里马拉杜项目
Miradorproject, Madrid

图13 Fig.13 马德里Celosia项目
Celocia project, Madrid

图14 Fig.14 利华大厦（SOM/1952年建成）
Lever House, SOM/completed in 1952

参考文献 References

1. 参见赖安·阿温特（2011年9月4日）"获得好工作的一条途径：更密集的城市"《纽约时报周日评论》，第SR6页，和格莱泽·爱德华（2011年5月8日）：以及爱德华·麦克马洪（2009年春季）"在规划中区分密集和高层建筑"《委员会期刊》第74期，第20页。See Ryan Avent, "One Path to Better Jobs: More Density in Cities", New York Times Sunday Review, 4 September 2011, p.SR6; and Edward McMahon, "Differentiating Between Density and High-Rise Buildings in Planning", Commissioners Journal, no. 74, Spring 2009, p. 20

2. 简·雅各布斯（1963年）《美国大城市的生与死》，纽约：古典书局。Jane Jacobs, The Death and Life of Great American Cities, New York, Vintage Books, 1963

3. 参见英德米特·基尔兰德、霍米·卡拉思以及迪帕克等人（2007年）"东亚复兴：经济增长理念之"（国际复兴开发银行/世界银行）和莎拉·威廉姆斯·戈德哈根"建筑：生活在高处"《新共和国周刊》.2012年5月18日。See Indermit Gilland, Homi Kharas with Deepak Bhattasali et al, An East Asian Renaissance: Ideas for Economic Growth, Washington DC, The International Bank for Reconstruction and Development/The World Bank, 2007; and Sarah Williams Goldhagen "Architecture: Living High in The New Republic", 18 May 2012, available at http://www.newrepublic.com/article/books-and-arts/magazine/103329/highrise-skyscraper-woha-gehry-pritzker-architecture-megalopolis (accessed May 2012).

4. 参见罗伯森·A·肯，"加拿大卡尔加里人行天桥：与美国市区体系的比较"《都市》第4卷第3期，1987年8月，第207-214页和罗伯森·A·肯（1993年）"市区规划员行人步道化战略：天桥与徒步购物中心"《美国规划协会期刊》，59：3，第361-370页。See Kent A. Robertson, 'Pedestrian Skywalks in Calgary, Canada: A Comparison with US Downtown Systems' in Cities, vol. 4, no. 3, August, 1987 pp. 207–214; and Kent A. Robertson, 'Pedestrianization Strategies for Downtown Planners: Skywalks Versus Pedestrian Malls', Journal of the American Planning Association, vol. 59, no. 3, 1993, pp. 361-370

5. 克里斯·亚伯，"面向新城市拓扑学"，《Planetizen》，2011年8月18日http://www.planetizen.com/node/50957 2012年6月检索。Chris Abel, 'The Vertical Garden City: Towards a New Urban Topology', Planetizen, http://www.planetizen.com/node/50957, 18 August 2011 (accessed June 2012)

6. 参见韩国交通研究院，"基础研究报告：超高层垂直城市的新交通系统" http://english.koti.re.kr/board/report/index.asp?mode=view&mcode=040100&code=research_report&cate=1&board_record=1417 2012年5月检索；Wan、圣西亚·W·S，（2007年）."天桥体系在香港中央商业区开发中所起的作用"，《美国旧金山美国地理学家学会年度会议》，2007年4月17—21日。参见社会科学研究网：http://ssrn.com/abstract=1699686或http://dx.doi.org/10.2139/ssrn.1699686 2012年5月检索；Iris Se Young Hwang，"从垂直延伸到容积城市的跳跃：关于紧凑可行的3D未来模型"，《2006年生态城市世界峰会会议录》，http://www.alchemicalnursery.org/index2.php?option=com_docman&task=doc_view&gid=187&Itemid=27. See The Korea Transport Institute, 'Basic Research Reports: New Transportation Systems for Super High-rise Vertical Cities', http://english.koti.re.kr/board/report/index.asp?mode=view&mcode=040100&code=research_report&cate=1&board_record=1417 (accessed May 2012); Sancia W.S. Wan, 'The Role of the Skywalk System in the Development of Hong Kong's Central Business District', Annual Meeting of the Association of American Geographers, San Francisco, United States, April 17–21, 2007, available at http://ssrn.com/abstract=1699686 or http://dx.doi.org/10.2139/ssrn.1699686 (accessed May 2012); and Iris Se Young Hwang, 'Leapfrogging from Vertical Sprawl to Volumetric City: A Study of Compact, Viable 3D Future Model Using Hong Kong', Eco-city World Summit Proceedings 2006, http://www.alchemicalnursery.org/index2.php?option=com_docman&task=doc_view&gid=187&Itemid=27 (accessed May 2012)

7. 约翰·斯马特·约翰（2005年）.高密度城市地下自动化高速公路（UAH）（PDF）.http://accelerating.org/articles/uahsframework.pdf. 2012年6月检索。John Smart (for Acceleration Studies Foundation), 'Underground Automated Highways (UAHs) for High-Density Cities', http://accelerating.org/articles/uahsframework.pdf, 2005 (accessed June 2012)

8. 气候保护合作部门（美国环境保护署办公室环境项目）"降低城市热岛效应：战略概要——绿色屋顶" http://www.epa.gov/hiri/resources/pdf/GreenRoofsCompendium.pdf 2012年6月检索Climate Protection Partnership Division (U.S. Environmental Protection Agency's Office of Atmospheric Programs), 'Reducing Urban Heat Islands: Compendium of Strategies – Green Roofs', http://www.epa.gov/hiri/resources/pdf/GreenRoofsCompendium.pdf, (date unknown), (accessed June 2012)

9. Isabelle Toland, '4 Unités LC: Fragments of a Radiant Dream' http://architectureinsights.com.au/media/uploads/resources/Isabelle_Toland_Report.pdf, 2012年5月检索。Isabelle Toland, '4 Unités LC: Fragments of a Radiant Dream', http://architectureinsights.com.au/media/uploads/resources/Isabelle_Toland_Report.pdf, 2001 (accessed May 2012)

10. Oliver Wainwright, "MVRDV的Celosia住宅" http://www.oliverwainwright.co.uk/2009/10/celosia-housing-by-mvrdv.html, 2012年5月检索 Oliver Wainwright, "Celosia Housing by MVRDV", http://www.oliverwainwright.co.uk/2009/10/celosia-housing-by-mvrdv.html, 2009 (accessed May 2012)

11. Albert Ferre和Tihamer Salijed，（2010年）《城市扩张总住宅选择》，巴塞罗纳Actar出版社 Albert Ferre and Tihamer Salijed, Total Housing: Alternatives to Urban Sprawl, Barcelona, Actar, 2010

巨构：召唤未来之梦
Megastructures: Evoking Dreams for the Future

主题 I：
立体发展方向3/4
Subject I:
Aspects of vertical development

黄一如
同济大学建筑与城市规划学院
副院长、教授
朱培栋
同济大学建筑学博士研究生

HUANG Yiru
Professor, Vice-Dean of College of Architecture and Urban Planning
Tongji University, Shanghai, China
ZHU Peidong
Ph.D Candidate of Architecture
Tongji University, Shanghai, China

近年来，增大居住密度已成为全球发展趋势。在中国，这已经成为一项急迫的现实需求。本文概述了中国在高密度居住环境领域所做的尝试及取得的经验，同时，也描述了唤起建筑领域巨型建筑历史之梦的一些相关背景、理论和技术研究。本文还讨论了巨型建筑存在的一些问题，比如社会结构组织、土地利用特点、室内环境控制及再生能源利用等等。

Recently, increasing residential density has been becoming a global development trend. As for China, it is already an urgent and realistic need. This paper outlines China's attempts and experience in the field of high-density residential environments. It also addresses some background on evoking the historic dream of megastructures in architecture, and the relevant studies in theory and technology. This paper also discusses problems of megastructures such as social-structure organisation, land-use features, indoor environment control, renewable energy utilisation, and others.

图1 Fig.1　香港鸟瞰 Aerial view of Hong Kong

在亚洲，城市人口高密度集中已造成了一系列社会和环境问题，如空气质量恶化、热岛效应、水污染、交通拥堵、生活质量下降、光污染、噪音污染等。其中，急剧增大的城市人口密度带来的问题尤为突出，且这种密度增大趋势似乎已经超出了过去传统的城市理论和现代居民区理论的经验范畴。[图1]

据估计，至2020年，将会有3.5亿人口从中国的农村地区迁往城镇地区。按现今的水平增长模式计算，至那时，城市所需用的土地面积将会是现今所使用的土地面积的两倍。考虑到有限的土地资源和急剧累积的人口数量，增大居住密度已成为中国社会的迫切需求和实际情况。因此，传统的建筑知识和经验在应对人类历史上最快速的、最大规模的城市化进程中显露出的一系列困难，中国急需新的方法。最近，不同的建筑实践或个人建筑师提出的理想化愿景都清晰地指向同一个终点：具超高容纳密度的城市巨构。

在发展成为服务于高密度生活的可能战略之前，巨构早已成为了建筑界千年来的终极梦想，这一点是可论证的。早在2000年以前，西汉的文学古籍中就有关于中天台故事的记载。魏国君主雄心勃勃地试图举全国之力造一座能达中天的高台[1]。最终在许绾的建议下，这一项目才废止。还有其他的例子，如众所周知的世界七大奇迹（包括吉萨金字塔、巴比伦空中花园、奥林匹亚的宙斯巨像以及亚历山大灯塔）都是当时的巨构。

自20世纪以来，关于巨构的建议和想法，从勒·柯布西耶的《阿尔及尔规划》（1932）到桢文彦关于巨型建筑的系统探讨（1961）[2]及至建筑电讯派和新陈代谢派的一系列雄心壮志，各种充满想象的巨构作品（20世纪60年代至70年代），都表明人类追求巨构的脚步几乎从未停止。近年来，随着过去20年里世界各地重新开始建造各种大型摩天大楼，这一雄心得以被再次证明。[图2至图12]

人类社会的集体潜意识表现中的英雄主义、复杂性及宏伟性和对自我超越和自我实现的永恒追求书写了巨构的悠久历史，给予巨构这一建筑之梦旺盛的活力。然而，另一方面，上升的建筑趋势或是建筑学进程并不能脱离外部社会条件的约束。纵观历史，与城市巨构直接相连的社会条件主要包括以下5个要点：土地缺乏、住房需求、资本积累、技术进步及乌托邦理想。可以证明的是，一旦同时出现两个或两个以上的要点，则马上会出现城市巨构相关的理论研究。

而如今的中国正处于这样的一个历史时机，五个要点同时出现：建筑用地匮乏，急速城市化带来的大量住房需求，在过去的30年间房地产持续发展累积了大量财富，建筑设计和建筑技术的进步，源于开发者和建筑师的乌托邦思想复兴。这些都对超级建筑赋予了与中国密切相关的重要意义。

历史机遇

1. 土地敏感和土地缺乏

中国的城市化趋势正处于跳跃式的发展，在已经实现城市化的地区，土地面临着人口增长带来的巨大压力。虽然中国的国土面积广阔，但60%的土地结构大都为山地、高原、山坡或沙漠，都不适用于建筑。除此之外，国家还设立了18亿亩[3]的耕地基线（12亿公顷）[4]来保障全国的粮食供应，这样一来，很大一部分土地就不能作为城市化建设用地。这意味着传统的水平、广泛发展模式已经不适用于目前的中国了[图13和图14]。因此，土地的缺乏和迫切的城市化需要为巨构提供了前提条件。巨构具有超高的可容纳密度和超高的容积率。

2. 大量住房建设需求

在城市化不断消耗土地的同时，中国正面临这史无前例的巨大的住房建设需求。可用土地的缺乏和土地价格的不断上涨导致仅仅依赖于建筑行业和市场的调整已经不足以应对目前的需要。因此，中国政府提出在未来5年要完成建设3600万套社会保障住房，到2011年[5]，解决1亿人的安置问题。

要在短时间内用现存的方法建成大量的住房存在局限。在现行的城市调节系统下，可行的城市住宅样式受控于指标和标准，而依照相关条例建成的常规高层或超高层建筑并不能达到超高的地积比率。4.0的建筑容积率似乎就已经是上限了。为解决这一问题，研究探索能达到并打破传统容积率限制的巨构（或其他建筑类型）已成为当务之急。

3. 资本积累

中国在20世纪80年代经历的加速大规模城市化进程毫无疑问地推动了房地产业的发展。国家试图通过市场经济来解决住房短缺问题[图15和图16]。拆迁、重建和"新式"城镇的建设使得过去的老城和农田迅速被现代小区替代。商品房政策积累了大量来自开发商和地方政府的资本。在过去的30年里，这一政策和房地产业的大规模发展相结合，把中国变成为世界上最大的建筑工地。这是中国除自2001年加入世贸组织后转变为"世界工厂"之外，又一项国家级的加速私人资产积累的事件。来自企业、地方政府以及中央政府的大量资金为建造巨型建筑提供了资金保障。

4. 技术进步和对技术的信任

随着中国的基础设施逐步完善，建筑能力逐渐提高以及建造摩天大楼技术日臻成熟，已经能够成功地在北京奥运会之前的一段时间内完成一系列大跨度、超大型建筑的建设，最近的上海世博会也是。[图17至图20]。通过预制装配式住宅建筑实践获得的工厂预加工及装配方法方面的进步、及在对生态无害的建筑实践中取得的巨大进步，都保障了建造巨构的技术安全。归功于建筑信息模型技术发展和数字建筑潮流，设计和控制复杂形式的能力达到了前所未有的水平。这大大增强了人们建造巨型建筑的信心，同时也使得建筑行业及其设计处理能够实现逐级发展。

5. 乌托邦思想复兴

经济的持续发展催生了人们对中国大楼在高度和规模上的英雄主义般的追求。据统计，目前，在中国大陆有超过152米（500英尺）的摩天大厦350座（非住宅型建筑和塔楼）。这一数字使得中国成为了世界上拥有超高建筑数量排名第一的国家。目前，中国在建的超高层建筑有287座（占世界在建超高层建筑总数的87%），此外，461座新的超高层建筑的设计工作也已经完成[6]。对高度的狂热追求最终将建设者们关于巨构的理想化的乌托邦思想变为现实。[图21至图24]

在这样的背景下，私营企业也启动了对巨构的研究和实践。比如，中国万通集团在2009年成立了一家名为"立体城"的专门公司[图25]，该公司与国内外多家在高密度住宅设计方面有经验的建筑工作室建立了联系。公司与这些工作室一道，进行了"垂直城市"课题的研究。万通集团还邀请AS+GG建筑工作室参与到一项"立体城市"项目中来——建造一座总建筑面积达600万平方米，每平方公顷可容约10万至15万人口的巨型建筑。

此外，中国远大集团（远大生态建筑）在2010年启动了"空中城"项目的第一阶段设计大赛。[图26]这一项目的最初目标是建成一座高666米共200层的城市巨构，可提供的总面积达124万平方米，容纳10万人。这一项目在2012年4月已进入了第二阶段，并且将设计方案修改为了建成高838米共220层的世界第一高楼。

巨构的优势

巨构之所以成为高密度住宅的终极形式，其原因可能在于它与常规形式相比较后，显现出在类型学上的五大优势。

1. 巨型的外表面

作为建筑，甚至是作为一座城来说，巨构始终持有巨大

勒·柯布西耶的阿尔及尔炮弹规划（1932）
Le Corbusier's Plan Obus for Algiers (1932)

康斯顿的新巴比伦（1959-74）Constant Nieuwenhuys' New Babylon (1959-74)

尤纳·弗里德曼的空间城市（1958）Yona Friedman's Spatial City (1958)

彼得·库克（建筑电讯）的插接城市
Plug-in City by Peter Cook (Archigram) (1964)

丹下健三的东京湾项目（1960）
Kenzo Tange's Tokyo Bay Project (1960)

"漂浮城市" "Floating City"

矶崎新的空中城市（1960）Arata Isozaki's Cities in the Air (1960)

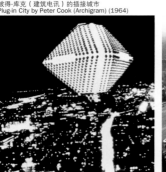

图2-12 Fig.2-12 荣久庵宪司的居住城市（1964）
Residential City by Kenji Ekuan

上海环球金融中心观景台景观View from Shanghai World Financial Center Observation Deck

菊竹清训的海洋城市（1960）Ocean City (1960s) by Kiyonori Kikutake

上海城市规划展览馆内的上海模型
Scale model of Shanghai at the Shanghai Urban Planning Exhibition Center

图13 Fig.13　上海郊区松江的住宅开发 Housing development in Songjiang, a suburb of Shanghai

图14 Fig.14　城市化之外开发程度较低的地区 View of less developed land beyond the urbanisation

的外表面。在当代生态敏感科技和新能源利用技术的支持下，巨构的巨大外表面为其在太阳能利用和采集雨水方面带来了决定性的优势；其不同于一般建筑的超高高度又为其带来了风能利用优势；以及潜在的生态和农业利益。

2. 可控制的内部环境

巨构在建设可控的内部环境的可能性上有优势。再生能源的使用，节能科技和新型材料的广泛应用，可建立起一个具有生态敏感基础的循环系统。这一系统将提供可控制的室内温度、气流、湿度和采光，在鼓励新的宏观尺度的气候控制思考的同时提高室内环境舒适度。

3. 紧凑的内部空间

与传统的城市风格和形式不同，紧凑的内部结构和空间是巨构的另一优点。巨构的内部成为一座紧凑的城，拥有最佳的占地面积和空间设计，且摒弃传统的运输系统。在这种超级城市中，适于徒步的布局和电气化的公共交通工具将成为首选。此外，强化、紧凑的内部空间也有助于提高能源效率，缩短运输距离，促进公共交流以及和谐的邻里关系。

4. 能源自治和社会自治

作为一座微型城市，巨构在能源和社会方面都提供了自治可能性。巨大的外表面产生的能源和再生能源科技的不断进步保障了能源自治的可行性。此外，依照德国和其他西方国家的污水处理和水循环经验来看，在不久的将来，巨型建筑就能够实现水自治。在中国，能容纳10万人的住宅乌托邦项目存在实现建筑内社会自治的可能性。中国城市广泛采用被形容为"次级政府，三方管理"的社会系统，规定设立街道办事处的人口规模为10万人[7]。在过去的50年里，作为城市里的政府终端机构，街道办事处及其下属的居委会在维护社会秩序方面扮演着重要的角色。这一历史系统使巨型大厦容纳的10万人实行自我管理，组建稳定的社会系统成为趋势。[图27和图28]

5. 非建筑用地的发展

上述四点优势形成了一个最终的决定性的特点：在一般定义为不适用于建筑的土地上建造建筑的可能性。巨大的外表面产生的稳定的再生能源流、独立的内部环境和社会机构、紧凑的内部空间以及有效的资源利用系统为建筑在土地开发方面提供了机会。无论现场的条件如何，无论是深深的矿井、陡峭的悬崖，还是广袤的沙漠甚或是在无边的海洋之上，巨型大厦都具有在保护环境的条件下，在不适用于建筑的土地上进行建设的优势。[图29]

范例所面临的挑战

俗话说凡事都有两面性。同样的，高密度住宅巨构在其具有一系列优势和高可实践性的同时也必然面临着一些挑战：

1. 挑战传统的城市文化

首先，巨构推翻了人们对于传统城市的看法。几千年来，横向发展始终是城市发展的主要方向，横向的移动，如步行、开车、坐公交等也是人们出行的主要方式。拥挤的广场，城市角落和公共空间诞生了多样化的城市生活和城市文化。然而，巨构的出现挑战了传统的城市文化和公共空间基础，横向街道和传统公共空间正在消失。高速电梯和垂直运输正逐渐成为城市新格局。这意味着城市移动的限制因素正从横向上的摩擦转变为纵向或三维空间上的地心引力。

其次，功能多样性可能会变得复杂，包括居住、工作、休闲、医疗、服务及其他方面。怎样将复杂的多样性全部涵盖其中？如何组织产业结构，安排合理就业？如何在紧凑、有限空间中实现社会化和隐私相平衡？这些疑问中都包含了巨大的问题。

这些问题并不是仅仅通过改善单一功能的超高多层建筑和酒店塔楼就能得到解决的，需要我们以革命性的态度思考并提出创新性的范例。

2. 经验缺乏与未知因素

随着今年来几个以建筑信息模型为基础的平台的应用，对诸如结构稳定性、抗震性及环境冲击性等因素的模拟已经不再是大问题。然而，心理层面上的模拟，即是说对人们对巨型大厦的可接受性的预测仍然是个未知数。成千上万人长期居住、工作在一个超大规模、高密度、封闭内部环境的空间中，对他们的人际关系将会产生怎样的影响是无法预测的。对居住在巨型建筑中的人们进行心理预测完全是一个未知数，需要进行长期的研究和试验。

3. 安全挑战

除此之外，在人类历史上从未出现过这样的巨构。大量人口、标志性形象和超高密度，导致巨型大厦成为恐怖袭击对象的风险非常高，这对居住在其中的人们来说会是巨大的威胁。安全管理、（灾难发生之前、发生之时和发生之后的）人员疏散以及一系列的安全问题都是巨型建筑面临的未知挑战。

总结

巨构好像拥有某种强大的魔力，或许近似于图腾一样，使我们为之着迷。然而，在当今的中国，它也有显著的可行性。不断的有范例转换为实体，其原因在于巨构通过高密度的住房所提供的价值、它在土地利用方面的优势以及它在模型逻辑方面的优点。

然而，另一方面，城市巨构看起来几乎是海市蜃楼。巨型建筑新的规模、巨大的人口量、超高的居住密度、复杂的功能以及未知的挑战使得无法适用从传统的摩天大楼中获取的传统经验。

全球人口超过70亿，中国城市化进程的急迫性使得中国必须采取进一步的行动来应对这些问题。在解决目前需要和展望未来发展的基础上，此刻或许就是中国建立关于巨构的理论体系和研究方法论的时机。巨型建筑——这一关于未来的最大的历史之梦正等着一步步成为现实。

In Asia, the super high-density accumulation of populations in urban areas has caused a number of social and environmental problems, such as the deterioration of air quality, the heat island effect, water pollution, traffic jams, the decline of quality of life, light and noise pollution, and so on. Above all, the problems brought by sharp increases of urban population density seem to have transcended the empirical range of traditional urban theories and modern residential area theories [Fig. 1].

A large volume of people is expected to migrate from rural to urban China in the next decade. The land required for urban areas will double, with calculations based on today's mode of horizontal growth. Given the limited land resources and the rapidly accumulating population, an increase of residential density has become an urgent need and reality of China. Traditional architectural knowledge and experience is proving insufficient to deal with the difficulties revealed by the fastest and largest process of urbanisation in human history. There is an immediate need for new methods. Recent utopian visions from different practices or individual architects are all clearly pointing to the same end: urban megastructures of super high density.

Before evolving into a possible strategy for super high-density living, megastructures had arguably already been an ultimate architectural dream for millennia. As early as 2,000 years ago, one piece of ancient Chinese

图15 Fig.15　典型的中国高密度住宅区-上海中远两湾城 A typically high-density residential district of China– Zhongyuan Double-bay City, Shanghai

图16 Fig.16　典型的中国高密度住宅区-上海汇龙花园 A typically high-density residential district of China – Huilong Garden, Shanghai

图17 Fig.17　2010上海世博中国馆（何镜堂）
China Pavilion at the 2010 Shanghai World Expo, designed by He Jingtang

图18 Fig.18　北京国际机场T3（福斯特及奥雅纳）
Beijing Capital International Airport, Terminal 3, designed by Foster + Partners and Arup

图19 Fig.19　北京国家体育场（赫尔佐格、德梅隆及艾未未和李兴刚）
Beijing National Stadium designed by Herzog, de Meuron with Ai Weiwei and Li Xinggang

图20 Fig.20　北京中央电视台大楼（OMA）
CCTV Headquarters, Beijing, designed by OMA

literature of the Western Han Dynasty had already recorded the story of the Middle Sky Tower. The king of Wei was ambitious and attempted to build a tower high enough to reach to middle of the sky using the whole country's resources. The project was finally abandoned upon the advice of Wan Xu. Other examples, such as the well-known Seven Wonders of the Ancient World (including the Great Pyramid of Giza, the Hanging Gardens of Babylon, the Statue of Zeus at Olympia, and the Lighthouse of Alexandria), were nothing but megastructures of their times.

Since the twentieth century, the proposals and thoughts – from Le Corbusier's Plan Obus for Algiers (1932), to the systematic discussion of megastructures by Fumihiko Maki (1961), to the series of ambitious and imaginative megastructures by Archigram and the Metabolists (1960s to 1970s) – show that there is hardly a pause in the dream in chasing the megastructure. This ambition has been proven again recently with the resumption of construction of giant skyscrapers worldwide during the last two decades [Fig. 2 to Fig. 12].

The heroism, complexity, and grandeur of the narrative in society's collective sub-consciousness, as well as the constant pursuit of self-transcendence and self-realisation, gives megastructures a long history and an exuberant vitality as an architectural dream. On the other hand, however, the rising of architectural trends or the progress of architecture can't be separated from the constraints set forth by external social conditions.

Throughout history, social conditions directly linked with urban megastructures have chiefly included five factors: deficiency of land, the need for housing, the accumulation of capital, the progress of technology, and the utopian ideal. Arguably, if two or more factors emerge at once, it is likely that new ideas and proposals for urban megastructures will arise.

Today, in a historic occasion, five of these factors are present in China: a lack of buildable land; a need for the mass construction of housing caused by rapid urbanisation; capital accumulation brought about by continuous real estate development in the last three decades; progress in design and construction technology; and the revival of utopian thoughts originating from developer and architect. These have given architectural megastructures a significant meaning that is relevant for China.

Historical Opportunities

1. Sensitivity and Deficiency of Land

China is leaping forward with new urbanisation trends, as land in urbanised areas comes under heavy stress from population increases. Although China's total land area is vast, 60 per cent of it is not suitable for building, being dominated by mountain, plateau, hillside, or desert. Besides this, a minimum of 1.8 billion mu of cultivated land (1.2 million ha) has been set aside to guarantee national food security, which also makes a huge proportion of land unavailable for urbanisation. It means the traditional development mode – horizontal and extensive – is no longer acceptable for China [Fig. 13 and Fig. 14]. Thus, the deficiency of land and the urgency of urbanisation provide a practical precondition for the megastructure, which holds a super-high density and a super-high plot ratio.

2. Need for Mass-construction Housing

While urbanisation continuously consumes land, China is also facing the unprecedented need for housing construction. The lack of available land and rising land prices have made it insufficient to solely rely on the housing industry and market regulation to deal with the needs. Therefore, the Chinese government proposed the goal of completing 36 million social housing apartments in five years in order to settle 100 million people by 2011.

The use of existing methods for the building of a huge volume of housing in a short time presented limitations. Under the current urban regulatory system, possible styles for urban residential buildings are dictated by an index and standards. The routine high or super high-rise buildings formed by those regulations can't achieve super-high plot ratios. A floor area ratio (FAR) of 4.0 seems to be the upper boundary. To address this, research and exploration of megastructures (or other building forms) that can reach beyond the conventional plot ratio limitation is a project of urgency.

3. Accumulation of Capital

The process of mass urbanisation that accelerated in China in the 1980s certainly raised the temperature of the real estate industry. The country tried to resolve the shortage of housing through the market economy [Fig. 15 and Fig. 16]. Demolition and reconstruction, and new town construction in 'fresh' styles, saw old cities and farmland rapidly replaced by modern gated communities.

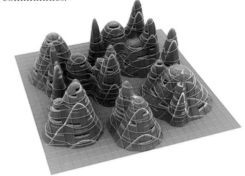

图21 Fig.21　中国山（MVRDV）MVRDV's China Hills project

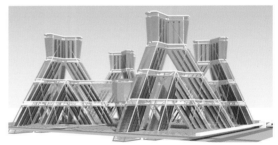

图22 Fig.22　千禧未来城（原田慎郎）Millennium Future City by Shizuro Harada

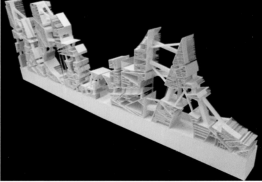

图23 Fig.23　空洞城市（鹈饲哲矢）Void City by Tetsuya Ukai

图24 Fig.24　超级细胞（UAA）Super Cell City by UAA

The commercial housing policy accumulated generous capital flow from real estate developers and local governments. Over the last 30 years, this, combined with the mass development of real estate, has turned China into the world's largest construction site. This is in addition to China's transformation into the 'world's factory' since joining the World Trade Organization in 2001, which accelerated the primitive capital accumulation at the national level. The huge amounts of money coming from companies, as well as local and central government, offers guaranteed funding for megastructures.

4. Progress and Confidence in Technology

The gradual maturity of China's infrastructure, construction abilities, and skill for skyscraper construction technology successfully enabled a great number of long-span and super-scale buildings during the period leading up to the Beijing Olympic Games and more recently the Shanghai World Expo [Fig. 17 to Fig. 20]. The advancement of factory prefabrication and assembling methods through prefabricated housing practice, as well as huge progress in ecologically friendly building practices, ensure the technical safety of megastructures. Due to BIM (building information modelling) technology and digital architecture trends, the ability to design and control complex forms has risen to an unprecedented level. This has greatly inspired people's confidence for the building of megastructures, which also enables the gradual development of the construction industry and its design processes.

5. Revival of Utopian Thoughts

The sustained and rapid development of the economy promoted the heroic pursuit of height and scale in buildings in China. According to statistics, there are currently 350 skyscrapers in mainland China (non-residential buildings and towers) that are above 152 meters (500 feet) in height. This number has already made China the top country in the worldwide ranking that measures the number of super high-rise buildings. Currently, there are 287 super high-rise buildings under construction in China (87 per cent of the world's total), and 461 more have had their planning completed. The reality and needs of urbanisation, the availability of capital, the progress of technology, and the fevered pursuit of height have finally materialised developers' idealised utopian thoughts about megastructures [Fig. 21 to Fig. 24].

Against this background, private enterprise has started its own research and practice on megastructures. For example, the Vantone Group of China set up a specialised company called 'Great City' in 2009 [Fig. 25] and associated with a number of domestic and international architecture studios that have experience in high-density residential design. Together, they carried out research on vertical cities. The Vantone Group also invited architecture studio AS+GG to participate on a project for a vertical city with a total floor area of six million square metres, which is able to accommodate 100,000 to 150,000 people in one hectare.

图25 Fig.25 立体城市（万通立体之城）GREAT City (Vantone Citylogic)

Moreover, China's Broad Group (Broad Sustainable Building) launched their design competition for phase I of a sky city in 2010 [Fig. 26]. This project initially aimed to build a 200-storey, 666-metre-high urban megastructure that would offer a total area of 1.24 million square metres and accommodate 100,000 people. The project progressed to phase II in April 2012, and to a revised proposal for the world's highest building with 220 storeys reaching 838 metres in height.

Advantages of Megastructures

Perhaps the reason that megastructures have become the ultimate strategy of form for high-density residences is related to the five advantages this typology can offer in contrast to any other routine form.

1. Immense Exterior Surface

As a building, or even as a city, a megastructure always possesses an immense exterior surface. With the support of contemporary ecologically sensitive technologies and new energy utilisation, some of the decisive advantages of the megastructure are the possibility of solar energy utilisation and rainwater collection brought by its enormous surface; the possibility of wind energy utilisation taken from the super high-rise structure; and the great potential ecological and agricultural benefits of the surface, among others.

2. Controllable Internal Environment

The megastructure holds the possibility of forming a controllable internal environment. The use of renewable energy, and a wide application of energy-saving technology and new materials, can establish a circulation system with an ecologically sensitive foundation. It will provide a controlled interior temperature, flow, humidity, and illumination, improving the degree of comfort while encouraging new, macro-scale thinking about climate control.

3. Compact Interior Space

Contrary to traditional urban styles and forms, a condensed interior structure and space is another merit of the megastructure. The interior of the urban megastructure becomes a compact city, which takes a highly optimised occupied area and spatial design to reject conventional private transportation. In this kind of mega-city, a pedestrian friendly layout and electric

图26 Fig.26 远大天空城市 Sky City by Broad Global

public vehicles will be the first choice. In addition, the intensified and compact interior space is helpful for raising energy efficiency, shortening distances travelled, and promoting public communication and neighbourhood life.

4. Energy Autonomy and Social Autonomy

As a miniature city, a megastructure offers possibilities for autonomy in terms of both energy and society. The energy that could be produced through the enormous surface and fast progress of renewable energy technology will ensure the possibility of energy autonomy. Furthermore, with the sewage treatment and water circulation experience of Germany and other Western countries, the megastructure will be able to realise water autonomy in the near future.

The 100,000-strong population of the residential utopia projects also hold the possibility of social autonomy in China. Chinese cities have widely adopted the social system described as 'secondary government, triple management', which stipulates that the population scale for a sub-district management office is 100,000. Over 50 years, as the final governmental agency in the city, the sub-district office and its subordinate neighbourhood committee have played an important role in keeping and managing social form. This historical system provides a tendency for a megastructure's 100,000-strong population to realise self-management and a stable social system [Fig. 27 and Fig. 28].

5. Development of Generally Unbuildable Land

The four advantages mentioned above lead to a final critical collective characteristic: the possibility of building on conventionally unbuildable land. A steady flow of renewable energy produced by the immense exterior surface, the self-contained interior environment and social structure, the compact interior space, and efficient resource-utilisation system give the megastructure opportunities in territorial development. No matter whether the site is in depths of a mine, on a precipitous cliff, in a vast desert, or even on the ocean, the megastructure has the advantage of being able to be constructed on otherwise unbuildable land under the precondition of environmental protection [Fig. 29].

Major Challenges of the Paradigm

It is said that a coin always has two sides. Similarly, while there are advantages and highly practical aspects to high-density residential megastructures, there are also several inevitable challenges:

1. Challenging the Traditional Urban Culture

Above all, the megastructure overturns the perception of traditional cities. For thousands of years, horizontal spread was always the main direction of urban development, and horizontal movements such as walking, driving private vehicles, and taking public transport were the basis. Varieties of city life and urban culture grew from crowded plazas, corners, and public spaces. However, the emergence of the megastructure challenges the basis of traditional urban culture and public space. While horizontal streets and conventional public spaces are perishing, express elevators and vertical transportation are becoming the new grid of the city, which means that the restrictive factor of urban movement is turning from horizontal friction to vertical or three-dimensional gravity.

Second, functional variety can be complex, including residential, work, leisure, medical, service, and others. How to contain the intricate diversity? How to organise the industrial structure and arrange for reasonable employment? How to balance socialisation and privacy in the compact and limited interior space? These questions involve great problems.

These challenges can not be solved simply by making improvements to the single-function super high rise such as office and hotel towers. They require us to think with a revolutionary attitude and create an innovative paradigm.

2. Lack of Experience and the Unknown

With the recent increase in the use of several BIM-based platforms, the simulation of factors such as structural durability, seismic behaviour, or environmental impact is no longer a big problem. However, psychological simulation – that is, predicting the level of human adaptability – is still an unknown. The impact on human

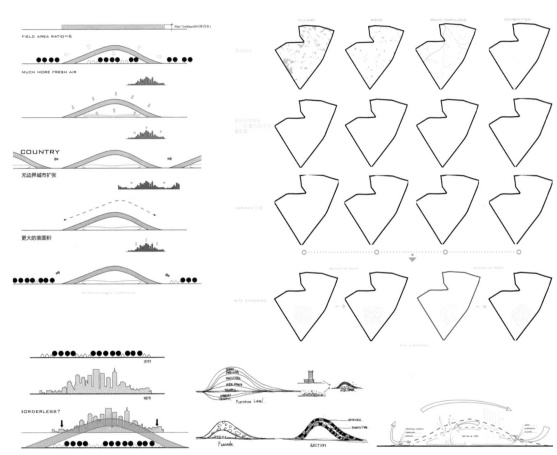

图27 Fig.27　无边际城市-巨大的外立面（同济大学）Boundless city – immense exterior surface by Tongji University

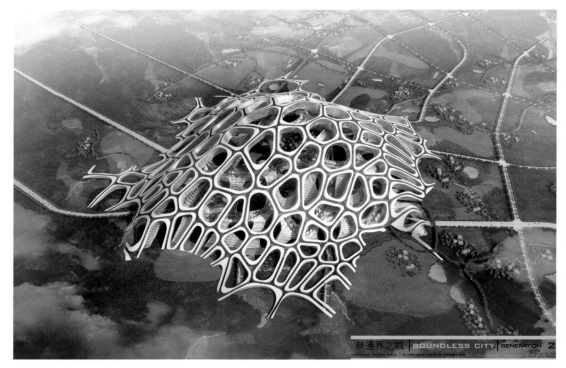

relationships of thousands of people continuously living and working in the same super-scale space, in super-high density, in a closed inner environment, cannot be predicted. The psychological perception of living in a megastructure is a total unknown that needs long-term research and trials.

3. The Challenge of Safety

There are no examples of such megastructures to be found in human history. The large population, symbolic image, and super-high density make the megastructure a high-risk target for terrorism, which could form a serious threat to its population. Regarding safety management, the evacuation of people (before, during, or after a disaster), and safety problems, megastructures present unknown challenges.

Conclusion

The megastructure seems to have a powerful magic, perhaps like a totem, that fascinates us continuously. However, it also presents distinct practical opportunities in China today. The paradigm is seeing translation into reality for the value it offers through high-density housing, its advantages of land utilisation, and its merits of pattern logic.

However, looking from the other side, the urban megastructure can seem almost like a mirage. The traditional experience acquired from conventional skyscrapers is not relevant because of the new scale, larger population, super high density, complex functions, and unknown challenges of the megastructure.

As the total population of earth exceeds seven billion, the urgent needs of Chinese urbanisation evidently require further action. Based on today's need and forecasts of the future, maybe now is the time to build a new theoretical system and research methodology for megastructures. History's biggest dream for the future is waiting to evolve.

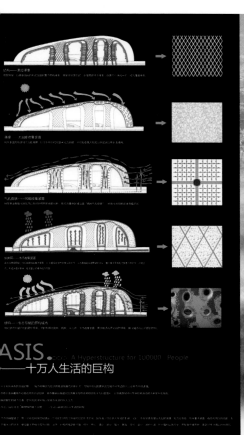

图28 Fig.28　绿舟-沙漠中的巨构（同济大学）Oasis – a megastructure in the desert by Tongjin University

图29 Fig.29　水平下的城市——未开发土地的发展
The City Under the Horizon – development on generally unbuildable land

参考文献 References

1 刘向（公元前24年），《刺奢第六·新序》，北京．中华书局出版社，200Liu Xiang (B.C. 24), Chapter 6 Cishe (刺奢第六), New Preface (新序), Beijing, Zhonghua Book Company, 2009
2 槇文彦（1964），《圣·路易斯集体形式的调查：华盛顿大学》。Fumihiko Maki, Investigations in the Collective Forms, St Louis, Washington University, 1964
3 亩是中国丈量面积的单位，一亩等于614.4平方米。"Mu" is a Chinese unit of measurement for area, and is equivalent to 614.4 square metres.
4 第十届全国人民代表大会第四次会议（2006）《第十一个五年计划中关于经济社会发展计划的指导方针》，中华人民共和国。4th Session of 10th National People's Congress, "Guidelines of Economic and Social development in 11th Five-Year Plan", People's Republic of China, 2006
5 第十一届全国人民代表大会第四次会议（2011）《第十二个五年计划中关于经济社会发展计划（草案）的指导方针》，中华人民共和国。4th Session of 11th National People's Congress, "Guidelines of Economic and Social development in 12th Five-Year Plan (Draft)", People"s Republic of China, 2011
6 《中国摩天大楼报告》，2011。www.motiancity.com（2011），Motian City, "Report of China's Skyscraper City", www.motiancity.com, 2011 (accessed July 2012)
7 中华人民共和国全国人民代表大会常务委员会（1954），《城市街道办事处组织条例》。The NPC Standing Committee of People's Republic of China, "City Sub-district Office Organization Regulations" 1954

主题 I:
立体发展方向4/4
Subject I:
Aspects of vertical development

朱文一
清华大学建筑学院教授
王辉
清华大学建筑学院教授

ZHU Wenyi
Professor, School of Architecture,
Tsinghua University
Beijing, China
WANG Hui
Professor, School of Architecture
Tsinghua University
Beijing, China

关于中国城市摩天大楼现状的研究
Study on the Situation of Skyscrapers in Chinese Cities

摩天大楼是城市立体发展的象征之一。最近几年,随着城市化快速发展,中国城市建起了大量的摩天大楼。这些摩天大楼是城市空间的重要节点,也提供了城市高密度发展的可能性。本文主要介绍中国城市摩天大楼的现状,包括其数量和分布情况。同时本文也包含城市摩天大楼的功能类型和作用分析,以及案例研究和开平碉楼的经验启示。

The skyscraper is one of the symbols of the vertical development of the city. In recent years, along with rapid urbanisation, a large number of skyscrapers have been built in Chinese cities. These skyscrapers can be important nodes in urban space and provide the possibility of high-density development for the city. This article introduces the situation of skyscrapers in Chinese cities, including their number and distribution. It also contains an analysis of functional types and the role of skyscrapers in the urban space, as well as case studies and the experience of Kaiping Diaolou.

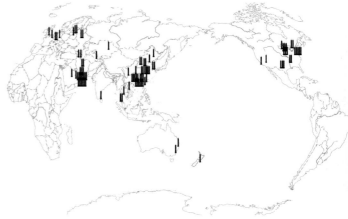

图1 Fig.1　世界摩天大楼分布 Distribution of skyscrapers around the world

亚洲的摩天大楼现状

基于现代建筑技术的发展,摩天大楼似乎已成为不可避免的建筑形式。摩天大楼最初源起于芝加哥,并在美国流行起来。随着全球化的发展,摩天大楼已成为全球城市建设的一种普遍现象。截至2012年6月,全世界共有105座超过300米高的大楼,其中大多数主要集中于中国、美国和阿拉伯联合酋长国三个国家,中国有34座,美国有19座,阿拉伯联合酋长国有18座,另外的34座分别位于其它23个国家。

亚洲是全世界人口最多的地方。为了应对高密度发展的需要,许多亚洲城市将建造摩天大楼作为解决办法。像一些已取得大规模城市化的西方国家一样,许多亚洲国家目前正经历着城市化快速发展。随着人口和人口密度的快速增长,亚洲城市数量和规模都发展迅速。如何发展高密度条件下的城市空间已成为亚洲城市开发研究的重点。

近几十年来,由于一些亚洲国家经济增长迅速,城市化发展加快,建设摩天大楼的速度也加快了,且大楼高度越来越高。摩天大楼不仅解决了亚洲城市的高密度问题,也创造出独特的"风景"。

今天,许多亚洲城市都建有摩天大楼[图1和图2]。全世界高度超过300米的摩天大楼中,亚洲城市就包含了71座,差不多占总数的70%。世界最高的前25座大楼中,亚洲城市包含19座,占总数的80%,其中中国有11座,西亚地区有5座,另外3座分别在东南亚国家。排名仅次其后的25座世界最高大楼中(排名26-50),亚洲有19座,其中中国占9座,阿拉伯联合酋长国占7座,其他国家占3座。再后面的25座大楼(排名51-75),亚洲有16座,其中中国占9座,阿拉伯联合酋长国占3座,其它国家占5座。前100名中排名最后的25座大楼中,亚洲占15座,其中中国占5座,阿拉伯联合酋长国占6座,另外4座位于其它国家。

中国的摩天大楼数量及分布

自1980年中国改革开放以来,中国正在建造的摩天大楼数量空前之多。摩天大楼在广州、深圳、上海、北京、香港、南京、天津和其它城市迅速崛起,建设规模和速度都很惊人。

经过30年高速、大规模的发展,越来越多的摩天大楼出现在中国,这一点也不令人惊奇。中国超过300米高的摩天大楼差不多占世界总数的1/3。数据显示,中国目前还有许多摩天大楼在建或正在规划。[1]毫无疑问,中国已成为摩天大楼王国,无数人生活和工作在300米高的大楼里。高度已成为中国城市建设的主流。

高于300米的摩天大楼主要位于中国的沿海城市[图3],集中在广州、深圳、香港(珠三角城市)以及上海和南京(长三角城市)。就中国摩天大楼的整体分布而言,东部地区比西部地区多,而且更多是聚集在东南沿海地区,这与中国不同地区的城市特色和经济发展水平相一致。

610米高的广州塔是中国最高的建筑,位于第二的是508米高的台北101。而位于上海的492米高的世界金融中心排名第三。排名第四的是香港484米高的世界贸易中心,而排名第五的是上海467.9米高的东方明珠电视塔。南京的紫峰大厦、深圳的京基100大厦、广州的国际金融中心(西塔)、上海的金茂大厦和天津的电视塔分别排名第六至第十。随着330米高的世界贸易中心刚进入世界300米高摩天大楼行列,北京中心商务区已初具规模。在发展完善的上海陆家嘴地区,有着一些著名的摩天大楼,包括东方明珠电视塔、420.5米高的金茂大厦和世界金融中心。在建的632米高的上海中心大厦将是中国最高的建筑。[2]

中国摩天大楼的功能类型

中国城市摩天大楼的功能用途可分为四种类型:观光塔、通信塔、多用途建筑和办公大厦[图4]。高度超过300米高的34座摩天大楼中,包括15座办公大楼、10座多用途建筑、5座通信塔和4座观光塔。考虑到建筑高度,最高的摩天大楼主要是多用途建筑,而低一点的摩天大楼多半是办公大楼。

34座摩天大楼中,有12座是2010年(摩天大楼增长最快的时候)后建立的[图5]。其它摩天大楼的建造时间都处于1989年至2009年的20年间。2000年至2009年的10年间,建造了10座摩天大楼。之前的10年(1989年至1999年),共建造了11座大楼。中国的第一座摩天大楼是于1989年建造的。摩天大楼的建造时间分布与改革开放30年的发展历程相似。特别是最近几年,随着经济的迅速发展以及大量城市建设项目的兴建,摩天大楼发展迅速。为了创造一个现代的城市环境,北京、深圳、上海、广州和其它大城市都纷纷出现多种类型的摩天大楼。

摩天大楼建设起因于城市经济、土地利用和科技的综合效应。在经济迅速发展的背景下,市场因素成为了摩天楼建设的驱动力。人们视建筑规模和建筑高度为现代化地标,因此,摩天大楼已成为城市现代化的象征之一[图6]。办公楼项目都趋于将高层楼建筑或摩天大楼作为其建筑形式。许多大型组织、知名企业和投资商对摩天楼建设都充满了兴趣,以便以此彰显各自的实力、自信和形象。

他们通过高大的建筑来炫耀自己的实力。建设摩天大楼不仅需要强大的经济实力作为支撑,而且还需要高级的建筑结构、建筑材料、建筑技术和结构设计作为保证和建设基础。

摩天大楼在城市空间中的作用

由于摩天楼规模、高度和形象对周边城市空间和城市的整体发展产生了影响,摩天大楼已成为一个城市层次的标志性建筑。摩天楼建设与中国城市空间模式的转变密切相关。最近几年,几乎所有拥有摩天大楼的城市都经历了快速的城市化过程,而摩天大楼在城市化过程中起了重要的作用。摩天大楼通常建立在新区中心,成为新城市区域的标志性建筑[图7]。面临土地不足的困境,高层建筑和摩天大楼成为新建工程的主要选择。设计和建造摩天大楼同样成为建立新城市中心的重要方法之一。摩天大楼对城市天际线有着重要的影响。摩天大楼的突出高度成为天际线的主要控制点。在上海浦东的陆家嘴地区,突出的天际线由东方明珠塔、国际会议中心和金茂大厦组成。天际线成为一个城市的标志性景观[图8]。另外,由于规模和体积巨大,摩天大楼对周围城市空间有着显著的影响。摩天大楼通常位于城市中心,特别是中央商务区的核心地区,这里聚集了各种高层建筑。由于其突出的建筑形式和势不可挡的高度,摩天大楼总是傲立于周围环境之中,显示出其强制性的影响和周围城市的全新形象。摩天大楼充实了周围区域的城市职能,进而增强了城市的活力[图9和图10]。

摩天大楼在解决城市环境中的高密度需求方面有一定的优势。它们提供了密度、强度、突出性和相关土地利用量最少的优势。摩天大楼能充分地展示出现代城市的魅力,成为城市经济实力的象征。另一方面,由于规模大,摩天大楼利用有限的区域吸引了大量的人和车,这将产生一系列的城市问题,如安全问题、能源问题和交通问题等。因此,有必要对摩天大楼建设作充分的可行性研究,避免盲目发展和建设。片面地追求最大和最好影响了合理、有序的城市发展。

案例研究及开平碉楼和丽江古城的经验启示

从建筑史的角度来看,较高的建筑也能创造良好的生活环境。在广东开平市,这里的早期建筑都很高,其空间形态是由高楼大厦和田园风光组成,高楼大厦和自然环境的整合形成了一种奇特的魅力。

根据历史记录,该地区在当年处于全盛时期时,有3000个开平碉楼。现在,其中的1833个碉楼位于开平市的15个城镇[3]。这些碉楼含有可居住的空间,是外国建筑技术和当地建筑文化的结合。因其数量多,形态漂亮,风格多样,开平碉楼已被列入世界文化遗产名录[图11和图12]。作为中国本土建筑的一种独特类型,开平碉楼展示出了其多文化的特征,包括中国传统建筑风格、西方建筑类型的整体形态和建筑细节。建造开平碉楼反映出当地人对外国文化的一种态度——吸取外国建筑文化的优势并将其用于本土需要,这表明当地人非常愿意吸收外来文化的优点并加以创新。开平碉楼打破了中国传统村庄的低矮天际线。自然景观加上单体建筑的纵向发展创造出一种与众不同的本土高密度建筑文化。

在大量的世界文化遗产中,丽江古城也是古城建筑的一个典型案例。不同于开平碉楼,丽江古城主要是低密度建筑形式。丽江古城位于云南省西北部,总面积约4平方公里,人口接近30 000人。丽江古城和开平碉楼能够给当代建筑和城市设计以启示。

这里提供了两种可使人们在较高密度的城市正常生活的方法,并且告诉我们该如何充分利用纵向的城市空间来安排在高密度城市环境中工作、生活、交流和娱乐的城市功能。如何从中学习并吸收良好的高层建筑艺术,以建造适于居住的摩天大楼?这些挑战不仅存在于困难的建筑技术和工程,还存在于建筑与环境的结合。

结论

最近,长沙宣布其将在望城县建造世界最高的摩天楼,引起了公众的关注。据说该摩天楼将建造为838米高,超出目前的世界纪录保持者迪拜塔10米。据报道说,作为集住宅、酒店、办公楼和学校为一体的220层综合大楼,其成本将超出40亿元。一则新闻报道的标题则辛辣地问道:"摩天大楼竞赛,地标还是浪费?[4]

一方面,作为地标,摩天大楼能够显示出一个城市的形象。另一方面,摩天大楼是巨大的赚钱机器,将城市其它项目的大量资源不断吸走。更糟糕的是,许多摩天楼项目不能达到收支平衡。同私营企业的推动一样,炙手可热的摩天大楼建造也是由政府通过减税、地价打折等方法而推动的,以便为本土的地标建设吸引投资。但是,这需要耗费15年甚至更长的时间才能通过租赁获取收益。因此,人们开始普遍关注最近摩天楼建设情况激增可能会导致严重的负面影响,尤其是在二线城市和三线城市。

在摩天大楼出现的早期,摩天大楼是否适于人类居住就已成为一个争议的话题。人们一直在做有关这方面的研究。尤其是"9·11事件"以来,耐火性能和"防恐"性能已成为摩天楼设计师新关注的一个问题。作为中央商务区地标,有关摩天大楼的争议同样成为了对中央商务区的争议,这加重了交通拥挤和信号拥挤等城市问题。摩天大楼对营造一个舒适的城市环境是有利还是有害?我们只能在不断涌现出的更多摩天大楼中找到答案。

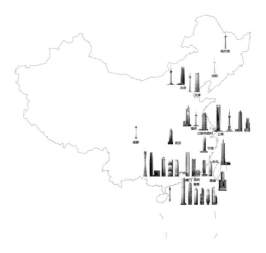

图3 Fig.3 中国摩天大楼分布 Distribution of skyscrapers in China

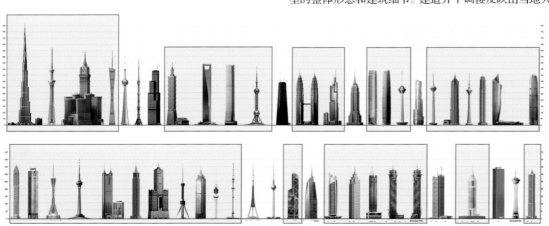

图2 Fig.2 亚洲摩天大楼占世界前50的比例 Proportion of Asian skyscrapers in the world's highest 50 skyscrapers

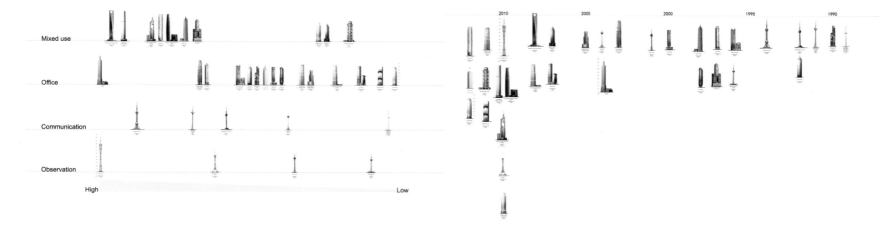

图4 Fig.4 中国城市摩天大楼功能分析 Functional analysis of skyscrapers in Chinese cities

图5 Fig.5 中国摩天大楼建成时间分析 Analysis of completion of skyscrapers in Chinese cities across time

Situation of Skyscrapers in Asia

The skyscraper has perhaps been an inevitable architectural form given the development of contemporary architectural technology. The construction of skyscrapers began in Chicago and became popular in the USA, and with the spread of globalisation they have become a phenomenon of global urban construction. Up to June 2012, there were 105 skyscrapers above 300 meters in height around the world. Most of them are concentrated in three countries: China, the United States, and the United Arab Emirates. There are 34 in China, 19 in the US, 18 in the United Arab Emirates, and 34 in 23 other countries.

Asia is the most populated continent in the world. A high number of Asian cities regard the construction of skyscrapers as one of the solutions to the need for high-density development. Unlike some Western countries, which have already achieved large-scale urbanisation, a significant number of Asian countries are currently experiencing a rapid urbanisation process. With rapid population growth and the concentration of people, Asian cities are quickly growing in number and size. How to develop urban space in high-density conditions has become the focus of urban development research in Asia. In recent decades, due to rapid economic growth and accelerating urbanisation in some Asian countries, the construction of skyscrapers has accelerated, with heights ever rising. Skyscrapers not only solve the high-density problem in Asian cities, but also create iconic "landscapes".

Of the world's 105 skyscrapers with heights above 300 metres, 71 are in Asia cities [Fig. 1 and Fig. 2]. This accounts for nearly 70 per cent of the total. Asian cities contain 19 of the world's top 25 tallest buildings, accounting for nearly 80 per cent of the total. These include 11 in China, five in Western Asia, and three in other Southeast Asian countries. Among the world's next 25 tallest buildings, Asia accounts for 19, including nine in China, seven in the UAE, and three in other countries. In the next bracket (51st-75th tallest buildings), Asia accounts for 16, including nine in China, three in the EAU, and five in other countries. In the 76th-100th-tallest bracket, Asia accounts for 15 buildings, including five in China, six in the UAE, and four in other countries.

Number and Distribution of Skyscrapers in China

Since the start of China's reform and the opening up of the 1980s, there has been an unprecedented volume of skyscrapers being constructed in the country. Skyscrapers are rapidly emerging in Guangzhou, Shenzhen, Shanghai, Beijing, Hong Kong, Nanjing, Tianjin, and other cities, with striking scales and speeds of construction.

It is not surprising that more and more skyscrapers are appearing in China with 30 years of high-speed and large-scale development. Skyscrapers in China with heights greater than 300 metres make up nearly one third of the total built in the world. Data shows that there are also many skyscrapers under construction or currently being planned in China. There is no doubt that China has become the kingdom of skyscrapers, with the largest population living and working above 300 metres. Height has become a fashion in China's urban construction.

Skyscrapers more than 300 metres in height are mainly located in China's coastal cities [Fig. 3], and are concentrated in Guangzhou, Shenzhen, and Hong Kong (cities of the Pearl River Delta), as well as Shanghai and Nanjing (cities of the Yangtze River Delta). In terms of the general distribution of skyscrapers in China, there are more in the eastern part than the western part and more aggregated in the southeast coastal area, which is consistent with urban characteristics and economic development levels of different regions in China.

The 610-metre-high Guangzhou Tower is currently the tallest building in China, followed by the 508-metre-high Taipei 101. The 492-metre-high World Financial Center in Shanghai ranks third in the list of tallest buildings. Fourth is the 484-metre-high World Trade Center in Hong Kong, and fifth is the 467.9-metre-high Oriental Pearl TV Tower in Shanghai. The Purple Peaks Building in Nanjing, the Jingji 100 Building in Shenzhen, the International Financial Center (West Tower) in Guangzhou, the Jinmao Tower in Shanghai, and TV Tower in Tianjin account for the sixth to tenth. The urban space of the CBD area in Beijing has begun to take shape, within which the third part of the World Trade Building with its 330-metre height has just entered the world's 300-meter-high skyscraper club. In Shanghai's Lujiazui district, which has been well developed, there are some famous skyscrapers including the Oriental

图6 Fig.6 中国摩天大楼高度前十名 The ten tallest skyscrapers in China

Pearl TV Tower, the 420.5-metre-high Jinmao Tower, and the World Financial Center. The 632-metre-high Shanghai Tower, which is under construction, will become the tallest building in China.

Functional Types of Skyscrapers in China

Skyscrapers in Chinese cities can be divided into four functional types: observation towers, communication towers, mixed-use towers, and office towers [Fig. 4]. In the 34 skyscrapers more than 300 meters high, there are 15 office buildings, ten mixed-use buildings, five communication towers, and four observation towers. Considering the height of the buildings, the tallest skyscrapers are mostly mixed-use and the majority of the shorter skyscrapers are office towers.

Among the 34 skyscrapers, 12 were built since 2010 when the fastest growth of the skyscraper typology appeared [Fig. 5]. The construction periods of the other skyscrapers are more uniformly distributed in the 20 years between 1989 and 2009. In the decade from 2000 to 2009, ten skyscrapers were built. In the preceding ten years, from 1990 to 1999, 11 were built. China's first skyscraper was built in 1989. The distribution of construction time echoes the 30 years of China's reform and development process. Especially in recent years, accompanied by the rapid growth of the economy and a high number of urban construction projects, skyscraper construction is happening very quickly. In order to create a modern urban environment, various types of skyscrapers are appearing in Beijing, Shenzhen, Shanghai, Guangzhou, and other big cities.

The construction of skyscrapers results from the combined effect of the urban economy, land use, and techniques. In the context of rapid economic development, market power has become the driving force for skyscraper construction. People consider building size and height to be the modern landmark of urban space, so the skyscraper has become one of the hallmarks of urban modernisation [Fig. 6].

Office construction projects tend to take the high-rise building or skyscraper as their architectural form. Many large organisations, well-known enterprises, and investors have a great interest in skyscraper construction to express their strength, confidence, and image. They want to flaunt their strength through the tall, large-volume skyscraper. The construction of the skyscraper requires not only a powerful economic strength as a foundation, but also advanced structures, materials, construction technology, and structural design as an assurance and basis for the construction.

Role of Skyscrapers in Urban Space

The great impact of its size, height, and image on the surrounding urban space and the whole urban development has made the skyscraper a city-level landmark. The construction of the skyscraper is closely related to the transformation of urban spatial patterns in Chinese cities. In recent years, almost all cities with skyscrapers experienced a rapid urbanisation process and skyscrapers played an important role in the process. Skyscrapers are often built in the centre of a new district and become iconic buildings of the urban area [Fig. 7]. Confronted with land scarcity, high-rise buildings and skyscrapers have become the main choice for new construction. The design and construction of skyscrapers has also become one of the important means of forming new city centres.

The skyscraper has a major impact on the urban skyline, becoming a control point due to its distinguishing height. In the Lujiazui area of Shanghai Pudong, the prominent skyline is composed of the Oriental Pearl Tower, International Convention Center, and Jin Mao Tower. The skyline becomes the city's iconic landscape [Fig. 8].

In addition, because of their large scale and significant volumes, skyscrapers have a substantial impact on the surrounding urban space. The skyscraper is often located in the urban centre, especially the CBD core area, where various kinds of high-rise buildings are concentrated. Skyscrapers demonstrate a brand new image for surrounding urban areas and enrich urban functions, thus enhancing the vitality of urban space [Fig. 9 and Fig. 10].

Skyscrapers have certain advantages in addressing the need for high density in the urban environment. They offer density and intensity, prominence, and relatively minimal land use. Skyscrapers can fully display the charm of the modern city and become symbols of the city's economic strength. On the other hand, because of their scale, skyscrapers attract a huge number of people and vehicles to a limited area. This will produce a range of urban problems – safety, energy, transportation, and so on. Therefore, full feasibility studies about the construction of skyscrapers should be undertaken. Blind development and construction should be avoided. The one-sided pursuit of the biggest and best affects the reasonable and orderly development of the city.

Case Study and Experience of Kaiping Watchtowers and Lijiang Old Town

From the perspective of architectural history, relatively high buildings can also make good living environments. In Kaiping, Guangdong, buildings had relatively high-rise features from a fairly early time. The spatial form there is composed of a scattered arrangement of tall buildings and pastoral scenery,

图7 Fig.7 广州、上海、香港深圳摩天大楼位置 Locations of skyscrapers in Guangzhou, Shanghai, Hong Kong and Shenzhen

图8 Fig.8 摩天大楼上海浦东天际线中的扮演重要角色 The important role of the skyscraper in the city skyline of Shanghai Pudong

with a unique charm resulting from the integration of the structures and the natural environment.

According to historical records, there were more than 3,000 Kaiping watchtowers in the area's heyday. Now there are 1,833 of them located in 15 towns in the Kaiping area.3 These towers contain liveable space, and are a combination of foreign construction techniques and local architectural culture. With their great number, exquisite form, and variety of styles, the Kaiping watchtowers have been added to the UNESCO World Heritage List [Fig. 11 and Fig. 12].

As a unique type of vernacular architecture in China, the towers exhibit multicultural characteristics including Chinese traditional local architectural styles and Western architectural styles in terms of the overall form and the architectural details. The creation of towers reflects an attitude of the local people towards foreign culture, which is to accept the advantages of foreign architectural culture and adapt it to local need. This shows the strong tendency in local people for absorption with innovation. The towers broke through the low skyline of the traditional Chinese village. The natural landscape environment and vertical development of single buildings created a distinctive vernacular high-density architectural culture in China.

Among the large number of world cultural heritage sites in China, the Old Town of Lijiang is also a classic example of ancient city construction. It is located in the northwest of Yunnan Province, with an area of about four square kilometres and a population of nearly 30,000 people. It features building forms with lower density than the Kaiping watchtowers.

The Old Town of Lijiang and the Kaiping watchtowers can both inspire contemporary architecture and urban design. They show two different ways of achieving relatively high density, taking full advantage of vertical urban space to arrange working, living, communicating, and entertaining areas. How can we learn from these examples and pass on the art of high-rise building to create a liveable urban space of skyscrapers? The challenge lies not only in the difficulty of building technology and engineering, but also in the relationship of the building with its environment.

Conclusion

Recently, Changsha's declaration that it plans to construct the tallest skyscraper in the world in its Wangcheng district has attracted much public attention. This future skyscraper is said to be 838 metres high, which exceeds the current world record-keeper (Dubai Tower) by ten metres. The cost of this 220-floor complex of residences, hotel, offices, and school is reported to be more than four billion yuan. The title of a news report poignantly asked: 'Skyscraper competition, landmark or waste?'

On one hand, as landmarks, skyscrapers can display the image of a city. On the other hand, skyscrapers are huge money-consuming machines that take away considerable resources from other projects in the city. Making matters worse, many skyscraper projects do not achieve long-term viability. The heated construction of skyscrapers is driven by the impetus of the private sector and also by the impetus of governments, which offer tax cuts, land price discounts, and so on to attract investment for the creation of local landmarks. However, it can take 15 years or even longer to achieve returns through rent. Therefore, there is widespread concern that the recent trend of skyscraper construction may lead to serious problems, especially in second-tier and third-tier cities. The connection of skyscrapers to urban problems such as crowdedness and traffic congestion in the CBD are also causes of controversy.

Thus, it appears that skyscrapers remain a contentious issue. New research is always being conducted into aspects of their design. For example, since September 11, the fire performance and "anti-terrorist" performance of skyscrapers have become pressing concerns for designers. Whether or not skyscrapers are helpful or harmful to the creation of a pleasant urban environment is yet to be determined, but China will certainly provide adequate examples for study.

图9 Fig.9 摩天大楼在深圳城市空间中扮演重要角色
The important role of the skyscraper for urban space in Shenzhen

图10 Fig.10 北京摩天大楼在城市空间的重要角色
The important role of the skyscraper for urban space in Beijing

图11 Fig.11 开平碉楼和民宅
Watchtowers and folk dwellings in Kaiping

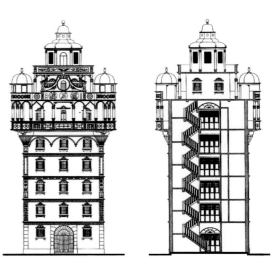

图12 Fig.12 典型开平碉楼的平面、立面、剖面
Plan, elevation and section of the typical watchtower in Kaiping

参考文献 References

1. http://skyscraperpage.com（访问时间：2012年4月9日）
http://skyscraperpage.com (accessed 9 April 2012).
2. http://skyscraperpage.com（访问时间：2012年4月9日）
http://skyscraperpage.com (accessed 9 April 2012).
3. "开平碉楼及民间住宅", 2007年, http://news.21cn.com/dushi/dspp/2007/06/28/3319141.shtml（访问时间：2011年11月28日）。
"Diaolou and folk dwellings in Kaiping", 2007, http://news.21cn.com/dushi/dspp/2007/06/28/3319141.shtml (accessed 28 November 2011).
4. "摩天大楼竞赛，地标还是浪费？", 北京青年报, 2012年6月27日, http://news.sjz.soufun.com/2012-06-27/7975673.htm（访问时间：2012年6月27日）。
"Skyscraper competition, landmark or waste?", Beijing Youth Daily, 27 June 2012, http://news.sjz.soufun.com/2012-06-27/7975673.htm (accessed 27 June 2012).

LEARNING FROM LA$ VEGA$

U-TT, Familia Perez y Felix Caraballo

Robert Venturi   **Denice Scott Brown**   **Steven Izenour**

主题 II：
人口老龄化空间和建筑老龄化 1/4
Subject II:
Spaces for the ageing of people, and the ageing of buildings

阿尔弗雷多·布利耶博格 & 休伯特·克伦纳
瑞士苏黎世瑞士联邦理工学院（ETH）
建筑与城市设计学院教授 / 委内瑞拉加拉加斯城市智库总监
与助理研究员伊拉娜·米尔纳和弗洛恩来·洛伦兹合作完成

Alfredo BRILLEMBOURG & Hubert KLUMPNER
Professors of Architecture and Urban Design, Swiss Federal Institute of Technology Zürich. Directors, Urban-Think Tank (U-TT), Caracas, Venezuela
Written in collaboration with research assistant Ilana Millner, and Florian Lorenz

城市老龄化：社会可持续性和参与性
Urban Ageing: Engendering Social Sustainability and Participation

在探索亚洲新兴大都市街道的过程中，阿尔弗雷多·布利耶博格和休伯特·克伦纳考虑了在当代城市化发展中贫民区和次级城市的重要性，着重于高密度方面。他们还提出了一个问题：成都能从拉斯维加斯学到什么？我们如何对单一功能商业高密集度的模式提出质疑？他们的议程将为边缘化人群创造解决方案，并提供社会性的可持续设计干预。

While exploring the streets of emerging Asian metropolises, Alfredo Brillembourg and Hubert Klumpner consider the importance of slums and secondary cities in the development of contemporary urbanism with a focus on the context of high density. They also pose the question: what can Chengdu learn from Las Vegas? How can we question the model of mono-functional, commercial high density? Their agenda is to create solutions for marginalised populations and to provide socially sustainable design interventions.

"城市并没有为人们作好准备，而人们也没有为城市作好准备。"
——《旧爱新欢一家亲》，委内瑞拉，加拉加斯

过去两年间，我们作为城市的探险者、流浪者，记录下了街道上不变而有限的生活体验。在集体讨论会期间，走在亚洲的街道上，我们产生了本文中提出的思路，寻找合适的值得考虑的情况，以及改造南半球城市贫民窟状况的想法。我们将目光投向这划分南北半球的政治赤道，穿过边界，不断地游走于世界各地。我们将有关全球城市景观的独特观点称之为"Gran Horizonte"。

非正式的亚洲

都市智囊团在苏黎世的联邦理工学院进行的工作包括许多亚洲城市深远结构转变的项目。在过去的十年间，我们已经见证了亚洲国家不断融合成为一个全球市场，在这个市场中，货物、服务、投资、人和知识不断地流动，这些资源都是可以互换的，不受距离或语言所影响。

为了开始我们有关当代亚洲城市的讨论，我们需要为我们的言论确定背景。将本文的第一部分主题定为"非正式的亚洲"是因为我们都接受但也拒绝有关城市的定义。我们所关注的亚洲城市的确不是根据任何传统而建立的，但是实际上它们是混乱的吗？它们缺乏组织吗？

如果在航摄照片上远看一下曼谷、雅加达、上海或成都，人们看到的是不规则伸展、如同根茎一般的形状；根本无法找到任何成序原则、一个清晰的开始和结尾，也没有办法将整个城市划分为可以理解的片区。我们清楚地看到如同根茎状延伸的城市就是亚洲超级都市的原型，并且代表了物质、空间、社会、心理和经济的一种新型交汇。这就转而要求用一种新的方式来构思和理解未确认的城市化类型。我们向二元对立的思考方式——正式对非正式、危险对安全、原始对现代、贫穷对富有——表示怀疑，但这些想法在当代城市化观念中根深蒂固，而这并不足以捕捉到那些新兴的城市形态。

我们打算要设计的城市类型尚未出现；没有人知道21世纪的城市将会是什么样子。但是，我们相信亚洲的城市化将成为21世纪城市演变的典型，并且亚洲城市为我们提供了很好的机会来研究我们谈及的新兴城市形态。本文下半部分分析的成都就是亚洲21世纪城市的最好例子。成都的基本条件对于中国其他城市来说很普遍，并且它具体表现了"发达"和"发展中"世界之间的交界——也可能是冲突。它让人们对20世纪的城市生活和发展模式提出质疑，并为复杂的城市问题提出了替代方法和解决方案。

我们总是认为世界是一个拥有70亿人口的快速生长的城市。根据联合国人类住区规划署2010年11月的世界城市报告，2010年世界城市占有率为50%，预计到2050年增长至70%。我们拥有边界交叠的庞大且复杂的大都市带，并且我们知道我们的研究必须集中在如何让密集的城市区域变得可以持续发展、能够创造价值、公平且适宜生活。我们作为建筑师的责任就是提高那些生活在标准以下人们的生活质量。由于城市发展得越来越复杂，行政和规划开始不起作用了，我们不得不面对一座没有经过深思熟虑的城市。这种无意识的"非正式城市"受到人类需求、活动和愿望的刺激，尚在创造过程之中，并且充斥着不断的变化和不稳定性；混合城市并没有遵循任何之前的模型或是普通的城市类型。

当北半球的人口在萎缩时，那些不发达国家被联合国人居署称为"贫民区"的区域正经历着人口快速增长——而"贫民区"这个词语起源于19世纪初伦敦贫苦工人的生活区域。今天，"贫困、自建城市社区"的名称——在格温德琳·赖特的描述中——随着区域的变化而变化：在土耳其称之为"geçekondu"、在中东称之为"compounds"、在巴西称之为"favelas"或"invasões"、在加拉加斯称之为"barrios"、在法语区称之为"bidonvilles"、在纳米比亚称之为"werften"、在菲律宾称之为"tondos"、在雅加达称之为

"kampung"等等。这份名单令人印象深刻,名称的差异为城区区域作为独立的国家提供了充分的理由:世界上七分之一的人口生活在贫民区——对于每个经济学家而言,这就是一个巨大的市场潜力。

世界各地生活在贫民区的10亿居民代表的人口与中国人口相差无几。据联合国人居署的统计资料,拉丁美洲和加勒比海地区23.5%的城市人口和东南亚地区31%的城市人口都生活在贫民区。雅加达、孟买、圣保罗、墨西哥城、拉各斯和上海等大都市内的贫民区代表了未来全球能源和基础设施需求的主要驱动力。

在本次讨论中,次级城市也同等重要。大多数次级城市的居民在50万到300万之间,但是在它们的国家或区域范围之外,这些城市通常都是不知名的。与欧洲和北美一两百年之前的城市发展相比,南半球的次级城市在接下来的几十年里将会经历大规模的扩张。由于城市及其人口增长,一切事物都会随着它们的增长而增长:财富和创造力,同时还有交通、犯罪行为、疾病和污染。

"正式"和"非正式"的城市建造程序已经相互交流并相互影响,每一种代表了同样恰当的城市现状。因此,我们必须提出一些基本但是撼动了我们偏见的问题:北欧的高科技基础设施如何能提升南半球的非正式生活水平?在过去十年间,交际广泛的社会活动家如何能提升北欧人的居住条件?我们怎样来解释南北双边对事实认知的不匹配,以便我们能将我们的调查结果转换为有成效的建设实践和可传递的知识?我们作为建筑师,如何能从以形式为导向的设计转而进入面向过程的设计?

忘却拉斯维加斯

1972年,丹尼斯·斯科特·布朗、罗布特·凡图理和史蒂文·艾泽努尔合著的颇具影响力的《向拉斯维加斯学习:被遗忘的建筑象征与寓意》出版。书中的第一章就清楚地阐述了一个有趣观点:"向现存的景观学习对于建筑师来说是一种革命方式。正如勒·柯布西耶在二十世纪二十年代所说的那样,并不是要拆了巴黎重新建设一个新的巴黎,而是以一种更加宽容的方式来进行;即质疑我们看待事物的方式。"斯科特·布朗、凡图理和艾泽努尔特别着眼于拉斯维加斯大道——他们称其为"最卓越的示范"——向那些不习惯"不含偏见地看待环境"建筑师,以及那些"宁愿改变现有环境,而不愿对现有的环境进行改善"的建筑师叫板。《向拉斯维加斯学习》成书于那个现代主义、极简主义、功能高于形式主义占统治地位的时代,让建筑师群体以一种新的视角融入老城区。

在成都的大街上闲逛时,《向拉斯维加斯学习》首次出版已经过了30年,我们偶然发现可比作臭名昭著的拉斯维加斯原型[图1、图2和图3]的天府广场延伸出了一条20公里长道路。在霓虹闪烁、贴满标牌的建筑中,我们回想起了斯科特·布朗、凡图理和艾泽努尔的话来,在混乱的冥想中寻找一些有用的指示。《向拉斯维加斯学习》提倡暂停判断,但是却不能无限期地暂停判断。我们必须提醒自己:标志性的建筑为中国文化所固有,但是在这个背景下,标志性建筑走向了极端。我们长期对亚洲新城市进行了细致的观察,并得出一个结论:必须减少使用拉斯维加斯模式[图4]。

北京奥运会和2010年上海世博会已成为过去,中国的关注转移到了挽救老城区遗迹上。在全球经济衰退带来影响时,密集的大型购物中心、商业大厦和庞大的公寓住宅区还在持续增长。中国已经有了非常多的经验,但仍需努力去保留住这些经验。预计在未来十年间将有3亿人口转移到城市,中国必须要重新考虑现行的有关社会住房的政策。与其运用西方有缺陷的城市化模式,为什么不参照中国建筑和城市组织的丰富遗产呢?中国必须保留住它的城中村,保护它的胡同、远离机动车回到有轨电车时代,并保留自行车、复兴街道的文化。

城市化和发展往往被以错误的角度来衡量——特别是在新兴的亚洲城市中更是如此。世界把太多目光集中在了财富的增长和楼层的高度上。我们还应以自主的幸福、社区稳定性、活跃的街道生活,以及与社会服务成功的融合来衡量发展状况。亚洲的大规模城市化项目存在巨大的潜力,但是我们必须利用一套新的理念和策略来应对这些挑战。为了为21世纪的新兴城市建设做好准备,本议题中提出的一些研究和项目试图从这些思维和行动中找到我们需要的工具。

将理论应用于实践来应对老龄化社会

为了应对现代城市带来的挑战,必须把设计师的基本角色扩展到改革策划者。老龄化社会的挑战在于如何将其重新定义为积极一体化和城市发展的机遇,而非将其当成是问题来看待。这一任务主动地抵制了不变的静止和还原的类型;它通过将现实结合"目的为导向"的社会设计策略注入实践中,形成处理一体化问题的新型城市模型。

这个方法拥有变革潜力,让人们更好地融合和参与到城市中。我们在委内瑞拉加拉加斯MetroCable项目已经实施了这种方法。它不仅是一个移动性的、无所不包的多模式系统,它还提供了解决多重人口并作为城市正增长催化剂的必要服务。附近的Avenida Lecuna也充当了解决这些问题的新建议的中心——它有一个灵活、不断发展的住房系统,包含基本的社会服务、为老龄人提供一个家,并与主要交通系统连接。我们的项目就是为边缘人群创造解决方案并提供让社会可持续发展的设计干预。

"The city is not prepared for the people, and the people are not prepared for the city."
– The Perez family; Caracas, Venezuela

In the last two years we have become urban explorers, wanderers of the city, recording the simultaneously timeless and time-bound experience of life in the street. The ideas presented in this essay were produced during collective brainstorming sessions while walking the streets of Asia, in search of appropriate conditions to consider and ideas for retrofitting the slum conditions of cities in the Global South. Our critical gaze is towards the political equator that divides North and South, and towards the horizon – the frontier of possibility – that constantly shifts as we move around the world. This particular view of the global urban panorama is what we call "Gran Horizonte".

Informal Asia

The work being conducted at Urban-Think Tank's chair at ETH in Zürich, Switzerland, includes projects that touch upon the profound structural transformation of many Asian cities. In the past decade we have witnessed the growing integration of Asian nations into a global market characterised by a flux of goods, services, investment, people, and knowledge, which are interchanged independent of the distance or language that separates one person from another. In order to begin our conversation about the contemporary Asian city, we need to establish a background for our observations. In calling the subject of the first part of this essay "Informal Asia", we are both embracing and rejecting the standard definition of the city. The cities of Asia on which we focus are, indeed, not made according to any conventional prescription; but are they in fact "disorderly"? Do they lack form?

If one looks at Bangkok, Jakarta, Shanghai, or Chengdu at a distance – in an aerial photograph – one sees sprawling, rhizome-like shapes; one searches in vain for an ordering principle, a clear beginning and end, for ways to separate the whole into comprehensible elements. It is clear to us that the rhizome-like expansion of these cities is the archetype of the Asian mega-city, and represents a new confluence of the physical, spatial, social, psychological, and economic. In turn, this requires new ways of conceiving and articulating the heretofore unacknowledged categories of urbanisation. We challenge the binary thinking – conventions of formal versus informal, danger versus safety, primitive versus modern, poverty versus wealth – that is ingrained in contemporary concepts of urbanism as simply inadequate for capturing these newly emerging urban forms.

图1 Fig.1 成都一景 View of Chengdu

图2 Fig.2 成都一景 View of Chengdu

图3 Fig.3 成都一景 View of Chengdu

The type of city we are going to design does not yet exist; nobody knows what this twenty-first-century city will look like. However, we believe that the urbanisation of Asia will become the model of city evolution for the twenty-first century, and that Asian cities present an excellent opportunity to study the newly emerging urban forms that we speak of. The city of Chengdu, which we analyse in the second half of this essay, is one prime example of the Asian twenty-first-century city. Chengdu's essential conditions are common to any Chinese city, and it embodies the interface – perhaps collision – between "developed" and "developing" worlds. It challenges one to question the patterns of late twentieth-century urban life and development, and to offer alternative approaches and solutions to complex urban issues.

We tend to think of the world as one rapidly growing city of seven billion. According to the UN-Habitat State of the World's Cities 2010/11 report, the world was 50% urban in 2010, and is predicted to be 70% urban by 2050. We have a huge, complex megalopolis with overlapping boundaries and we know that our research must be centred on making this dense urban area a sustainable, productive, fair, and inclusive place to live. It is our obligation as architects to ameliorate the quality of life for all humans living in substandard settlements worldwide. As cities grow increasingly complex, administration and planning begin to malfunction. We are left with a city that has formed without deliberation. The unintended "Informal City" is an ongoing creation, fuelled by the pulse of human needs, activities, and aspirations, and framed by constant changes and insecurities; the hybrid city does not follow any prefixed model or common city type.

While the populations in the northern hemisphere are shrinking, immense population growth is taking place in underdeveloped countries within areas that the United Nations agency UN-Habitat declares to be "slums" – a term originating in miserable workers' districts in London at the beginning of the nineteenth century. Today, the names for "impoverished, self-built urban communities" – a description by Gwendolyn Wright – vary by region: geçekondu in Turkey, compounds in the Middle East, favelas or invasões in Brasil, barrios in Caracas, bidonvilles in the French-speaking world, werften in Namibia, tondos in the Philippines, kampung in Jakarta, and so on. This list is impressive, and the differentiation by name likens urban territories to independent countries for good reason: one-seventh of the total world population lives in a slum – an immense market potential for every economist.

The one billion slum dwellers worldwide represent a population comparable to that of China. According to UN-Habitat, as of 2010, 23.5 per cent of the urban population of Latin America and the Caribbean, and 31 per cent of the urban population of Southeast Asia, were living in slums. Slums inside megacities like Jakarta, Mumbai, Sao Paulo, Mexico City, Lagos, and Shanghai represent the key drivers for the future global demand on energy and infrastructure. Secondary cities are of equal importance in this conversation. Most secondary cities have between 500,000 and three million inhabitants, but are often unknown outside of their national or regional context. Secondary cities in the Global South will undergo massive expansions in the next few decades, comparable to city growth in Europe and North America one-to-two hundred years ago. As cities and their populations grow, everything else grows with them: wealth and creativity, as well as traffic, criminality, disease, and pollution.

"Formal" and "informal" city-making procedures already inform and influence each other, each representing equally valid urban realities. Thus we must ask fundamental questions that shake up our preconceptions: How can Northern European high-tech infrastructure upgrade informal living in the Global South? How can the communicative activist competence gained during the past decade upgrade the living conditions in the North? How can we translate mismatching perceptions of reality on both sides, so that we turn our findings into fruitful building practice and transferrable knowledge? How can we, as architects, move from form-oriented to process-oriented design?

Un-learning Las Vegas

1972 marked the publication of Denise Scott Brown, Robert Venturi, and Steven Izenour's influential book Learning From Las Vegas: The Forgotten Symbolism of Architectural Form. One of the book's seminal points was clearly stated in its first chapter: "Learning from the existing landscape is a way of being revolutionary for an architect. Not the obvious way, which is to tear down Paris and begin again, as Le Corbusier suggested in the 1920s, but another, more tolerant way; that is, to question how we look at things." In particular, Scott Brown, Venturi, and Izenour looked at the Las Vegas Strip – what they called "the example par excellence" – to challenge architects unaccustomed to "looking nonjudgmentally at the environment," and who "have preferred to change the existing environment rather than enhance what is there." Written at a time when modern, minimalist, form-follows-function mentalities were dominant, Learning from Las Vegas pushed the architectural community to incorporate the old city in their visions of the new.

While wandering the streets of Chengdu, over 30 years after the initial publication of Learning from Las Vegas, we happened upon the construction of a 20-kilometre strip leading out from Mao Plaza that could only be likened to the infamous archetype of Las Vegas [Fig. 1, Fig. 2, and Fig. 3]. Among flashing lights, and signage glowing on every building, we thought back to the words of Scott Brown, Venturi, and Izenour, looking for some productive guidance in the midst of a chaotic meditation. Learning from Las Vegas advocates a suspension of judgment, but judgment cannot be suspended indefinitely. Symbolic architecture is inherent to Chinese culture, we had to remind ourselves, but in this context, it has been taken to the extreme. We have looked long and hard at the new Asian city and we conclude: the adoption of the Las Vegas model must be curtailed [Fig. 4].

Now that the Beijing Olympics and the 2010 Shanghai World Expo are but a memory, the spotlight in China is moving toward attempts to save the vestiges of the old city. While the global economic slowdown has had its impact, a profusion of shopping malls, commercial high rises, and gargantuan apartment blocks for the urban poor continue to rise in

图4 Fig.4《向拉维加学习》Learning from La Vega

cities across the country. China still offers a breadth of experiences, but it must work to preserve them. With an anticipated 300 million people moving to cities in the next ten years, China will have to rethink its current policy on social housing. Instead of appropriating flawed models of Western urbanisation, why not reference the rich heritage of Chinese construction and urban organisation? China must save its urban villages, protect its heritage of hutongs, move away from the car to the tram, keep the bicycle, and revive the culture of the street.

Urbanisation and development are measured in flawed terms – particularly in emerging Asian cities. The world focuses too much on the growth of wealth and the height of buildings. Development should also be measured in self-directed happiness, community stability, active street life, and successful integration of social services. There is great potential for large-scale urbanisation projects across Asia, but we must approach them with a new set of concepts and strategies. Some of the research and projects presented in this issue attempt to derive these needed tools of thought and action, in preparation for the emerging cities of the twenty-first century.

Applying Theory to Practice for the Ageing Society

To face the challenges of the contemporary city, it is necessary to expand the fundamental role of the designer into an animator of change – an agent provocateur. The challenge of the ageing society is to redefine it not as a problem but as a chance for positive integration and urban growth. This role actively counters the perpetual multiplication of static and reductive typologies; it injects practice with 'purpose-oriented' social design strategies by engaging with the realities on the ground to form new urban models that address integration. This methodology has the transformative potential to engender broad demographic integration and participation in the urban context. This has been put into practice with our MetroCable project in Caracas, Venezuela. Not only does it serve as an inclusive, multimodal system of mobility, but it also provides necessary services that address multiple demographics and act as urban catalysts for positive growth. The nearby Avenida Lecuna also serves as an axis of new proposals that address these issues – with a flexible, growing house containing essential social services, a home for ageing residents, and connections to major public transportation. Our agenda is to create solutions for marginalised populations and to provide socially sustainable design interventions.

高密度环境中适合老龄化人口的城市空间
Urban Space for an Ageing Population in a High-density Environment

主题 II：
人口老龄化空间和建筑老龄化2/4
Subject II:
Spaces for the ageing of people, and the ageing of buildings

CHO Im Sik
新加坡国立大学设计与环境学院建筑系助理教授
Zdravko TRIVIC
新加坡国立大学设计与环境学院亚洲城市可持续发展研究中心研究人员

CHO Im Sik
Assistant Professor, Department of Architecture, School of Design and Environment National University of Singapore
Zdravko TRIVIC
Research Fellow, Centre for Sustainable Asian Cities (CSAC), School of Design and Environment National University of Singapore

在对全球惊人的城市化速度、城市人口激增和城市混合发展作出的回应中，本论文提出了理解、分类、评估和指导新城市空间设计及规划的框架，并重点强调高密度环境。框架将设计（硬件）、使用（软件）和操作（组织件）方面看作主要的镜头，目的在于识别城市环境的密度和质量之间的复杂关系。通过这些主要的角度来调查塑造城市空间性能的关键属性。在对新加坡老年人口增加作出的回应中，对初始城市空间框架进行了升级，以评估和过滤城市空间为高密度环境中老龄化社会问题提供支持的关键能力。在重构城市空间的作用和能力的同时，升级后的框架对适合所有年龄阶段的支持性环境和治愈环境进行整体分析，从而鼓励去除老年人"积极养老"和"居家养老"的烙印。

In response to the dramatic speed of urbanisation, the boom in urban populations, and hybrid urban development globally, this paper proposes a framework to understand, categorise, evaluate, and guide the design and planning of new urban spaces with an accent on high-density conditions. With an objective to discern complex relationships between density and quality of urban environments, the framework recognises design (hardware), use (software), and operational (orgware) aspects as key lenses through which to investigate critical attributes that shape the performance of urban spaces. In response to a growing ageing population in Singapore, the initial Urban Space Framework is upgraded to evaluate and filter critical abilities of urban spaces to support the issues of the ageing society in high-density contexts. While reconceptualising the role and capabilities of urban spaces, the upgraded framework adopts a holistic approach to supportive and healing environments for all ages, encouraging de-stigmatisation of the elderly, "active ageing", and "ageing-in-place".

图1 Fig.1　城市空间框架-硬件、软件及组织件
Urban Space Framework – hardware, software and orgware components

高密度环境中的城市空间

本世纪之交的特点为全球城市人口急剧增加和城市发展极速飞跃。自2007年起，一半以上的世界人口居住在城市，而且预计在接下来的35年时间里，这一数字将增加到三分之二[1]。预计在2000年和2030年之间，亚洲的城市人口将从13.6亿增加到26.4亿[2]。

在这些城市化和人口变化趋势以及增加对环境和社会可持续规划进行探索的情况下，我们理解、构想、设计和利用城市空间的方法再一次受到挑战。一旦密度增加，空间便会成为一种珍贵商品，导致充满活力和多样化的用户群体之间发生激烈竞争、矛盾和谈判。因此，发现新方法为所有用户群体创造（重建）适合密集的城市环境，但也充满了活力以及具有环境和社会可持续性的城市空间将成为一项主要的挑战。

城市空间框架

过去的研究主要集中在熟悉的城市空间（比如，广场、购物中心、街道或公园）模型上。关于当今出现的越来越多的混合和密集城市空间类型，我们了解的情况仍然是非常的少。现有的文献和研究并未提供针对理解和评估高密度以及高强度环境中的新城市空间而有目的地制定的特定值或标准。在这些新情况下，将有必要考虑城市空间性质中所存在的差异，而非与现状的差异程度，包括推测尺度，主要的重点在于空间类型之间的复杂关系、计划以及利用和管理方法。

作为回应，我们研究的主要目的在于根据影响空间性能的关键标准来区别城市公开空间的密度和质量之间的关系。这一目的已通过制定城市空间框架予以实现。之后的目的在于派生一种用于编目、分类、评估、分析和推测混合城市空间类型、条件和性能的系统。

方法

根据吉尔[3][4][5]、卡莫纳[6]尤其是沙夫托[7]的城市设计理论及研究，提出了初始概念研究框架，并记录和分析了高密度环境中的26个新加坡城市空间和25个国际城市空间。案例研究纪录涉及到有条理的第一人称观察资料，包括使用观察清单对空间特点及强度的现场记录、制图、绘图、摄影、文本描述和偶尔采访，以及对二手资料的综合评价。

框架

一组初步案例研究分析起到了改善研究框架的作用，从而形成了最初的城市空间框架。最初框架承认了三个并存的主要组成部分。这三个主要组成部分构成了城市空间的性质和性能，即：硬件、软件和组织件。硬件是指城市空间的设计价值——城市空间的物理性质和几何性质。软件将空间的活动和社会感知价值结合在一起。组织件是指公共空间的操作和管理方面,而且经常与硬件组成部分和软件组成部分相连接或重叠[图1]。城市空间框架由分类系统和评估系统组成。这两个系统为识别城市空间关键属性和评估其性能的描述机制。

分类系统

分类系统由很多用于描述城市空间并将其归类为默认的主要类型、次要类型和混合类型的描述符和标签。根据主要描述符——S1:主要用途——系统识别了五个默认的城市空间类型,即住宅区、娱乐区里、城市中心、混合集成开发区和基础设施交通引导区的城市空间。

评估系统

评估系统由92项标准组成,用于描述和评估城市空间的性能[图2]。这92项标准按等级被分组为46个评估物、13项属性和5个城市价值。

评分系统

评分系统允许根据评估清单来评估城市空间,包括在评估系统中提出的所有标准。如果空间满足了一项标准,则空间的得分为"1"分。最终得分为所有已满足的标准的总和,并构成了城市空间的整体价值,用百分比表示。

圆形图

用两个圆形图来呈现我们的研究得出的分数——城市空间价值图(同时也用作评估清单)和城市空间价值饼图[图3]。圆形图分为三个部分,每个部分代表着城市空间的一个组成部分(硬件、软件和组织件)。每部分由条纹组成,每个条纹代表着一项评估标准。如果空间满足了标准,则给条纹涂上颜色。城市空间价值图显示特定空间的整体性能,而城市空间价值饼图显示的是空间每个组成部分的性能。

分析

利用一致性分析来调查所有相同类型城市空间的每项标准的得分。根据每项标准的平均分偏差来确定标准的等级系统。该等级系统识别:(a)基本标准——所有相同类型的空间最常满足的标准;(b)难以达到的标准——所有相同类型的空间几乎或从未满足过;(c)期望标准——所有空间未一致满足的标准;以及(d)关键标准——所有相同类型的城市空间多数未一致满足的标准。因为关键标准最可行,而且改善城市空间整体性能的纠正潜力最高,所以重点在于关键标准。

综合——城市空间分析工具(TUSA)

综合文件、分类、评估和分析开发出了一种集成计算工具——城市空间分析工具(TUSA)。除了用作交互目录之外,城市空间分析工具还主要被用作分析工具和推测工具,从而提供富有成效的研究、理解和指导城市空间设计过程的方法。

人口老龄化与城市空间——人口趋势与举措

根据对全球人口的预测,2025年65岁以上的老龄人口将从目前的3.9亿增加到8亿,占全球总人口的10%。亚洲65岁以上的老龄人口将飙升314%,从2000年的2.07亿增至2050年的8.57亿,占亚洲总人口的18%。[8]

新加坡的老龄人口增长速度在亚洲各国中排名前列。根据部际委员会《人口老龄化报告》[9]的预测,到2030年,新加坡65岁以上老龄人口占全国总人口的比例将从1999年的7.3%增加到19%。换句话说,目前新加坡每12人中有一名老年人,而到2030年,将达到每5个人中就有一名老年人,增加将近三倍。此外,未来老龄人口的受教育程度更高,更加独立,工作年限也将延长,因此,他们要比现在的老年人有着更多的需求与渴望。

"积极养老"与"居家养老"

认识到人口老龄化趋势蕴含的的潜力(而今仅仅是注重限制),1990年代,世界卫生组织开始使用"积极老龄化"这个术语,并开展了均衡、健康与快乐变老的项目。人类变老的过程被定义为"在人类变老的过程中为了提高生活质量而最优化地利用健康、参与和安全的机会"。世界卫生组织[10]确定了"积极变老"的六组关键因素,即社会,经济,健康与社会服务,行为,个人与物质环境,文化与性别则是跨领域的因素。

目前,"积极养老"已经成为全世界普遍接受和应用的主流概念,当然也包括新加坡,主要内涵还包括改善建成环境,改变人们对老年人的固有偏见与歧视,并开展社会融合。因此,我们面临的主要任务是鼓励"积极老龄化",通过提供全面的配套建成环境,打造老年人独立与健康生活的能力。

尽管从世界范围看老年人单独居住的趋势日益上升,但是家庭养老仍是亚洲的主要养老方式,新加坡也不例外。大约85%的新加坡老龄人口与至少一个子女共同生活。因此,同"积极养老"一样,"居家养老"也是新加坡公屋制度遵循的一个主要原则,其目的就是为了帮助老年人在自己的住宅中度过晚年[11]。老有所居的定义是,"不管年龄、收入与能力水平如何,都有能力安全、自给并舒适地生活在自己的住宅和社区内"[12]。

在"积极养老"和"居家养老"原则的指导下,新加坡制定了包含三方面因素的全面概念,即"关爱意识"、"软件"与"硬件"。"关爱意识"指的是促使社会价值观向针对变老过程和老年人的积极态度和关爱进行转型。软件包括为老龄人口中各不同群体提供的社区基础设施、活动项目与看护服务。[13]"硬件"则包括建成环境的各个方面。在这一战略框架下,各地开展了诸多项目,包括提供老年友好型住房,可进入性强的无障碍环境,价格适中的医疗与老年看护,以及宣传积极生活方式和福祉的各种活动。[14]

高密度环境中适合老龄化人口的城市空间

基于积极老龄化框架和在22个国家和33个城市中进行的调查,世界卫生组织于2007年为城市规划师和设计师首次发布了《全球关爱老人城市指南》[15],该指南确定了关爱老人城市环境的关键物理、社会和服务属性。这其中确定了八项关键属性,即,(1)户外空间和建筑;(2)交通;(3)住房;(4)社会参与;(5)尊重和社会融入;(6)公民参与和就业;(7)交流和信息;(8)社区支持和卫生服务[图4]。

为了满足由于高密度发展以及日益多样化的人口而引起的更高的要求、特殊性和复杂性,城市空间分析工具和城市空间框架的初始版本增加了一项特征,即使研究人员能够在更具体的方面评估城市空间性能的"过滤器"。更准确地说,世界卫生组织指南提供的建议用于缩小最初框架中提议的标准,并为老龄人口开发升级城市空间框架。

除了世界卫生组织关爱老人城市发展的各方面,升级框架保留了高密度条件下公共空间中的成功老龄化相关的附加标准。其中一些是指:空间层次内和之间的良好视线(因为这有助于理解空间构型和提高导向标识和安全性);自行车设施(有助于提高积极和健康的生活方式以及无车式发展);视觉地标和活动焦点(因为它们有助于创造多代共同生活、安全和空间识别度);多样性和绿化(因为它支持娱乐性使用,活跃感知并具有健康恢复功能);多种树荫和阳光条件(这会增加舒适感和选择);正式和非正式的座位和互动元素的类型和多样性(因为这创造了与空间之间以及在空间内的社会和多感官互动点);以及环境照明(有助于美感和导向标识,提高延长时间的使用)。在增加修改以便更好地论述老年人口时,新框架适用于所有的空间类型和所有用户群,包括儿童和成年工作者,应牢记每个人的年龄。

应用和调查结果

正如本文前面部分所述,选择的所有本地和国际公共空间都会在升级框架的基础上重新评估,并且使用城市空间分析(TUSA)工具来分析。最初的调查结果显示,大多数的案例研究中硬件部分都运作很好,而软件部分的分数最低。51处空间中有17处达到了整体老龄化城市空间价值中80%及以上的分数。但是,主要的目标是符合一套标准,该标准对成功老龄化过程和选择的分类标签——类型学的整体城市空间性能都十分重要。而这种分析是针对所有默认的城市空间类型,关于住宅区和城市中心(广场)中的城市空间,本文讨论了有关重要标准和推荐改善区域的重要调查结果,着重于一些最好的国际和当地的实例。

重要标准和需改善的区域——主要居民区的城市空间

[图5]通过一致性分析以图解的方式总结了主要居民区的

components	urban values	attributes	evaluators
HARDWARE	SPATIAL VALUE	01. Accessibility (Pedestrian)	(4)
		02. Connectivity	(3)
		03. Mobility Means	(4)
	NODAL VALUE	04. Legibility & Edges	(4)
		05. Spatial Variety	(2)
	ENVIRONMENTAL VALUE	06. Environmentally Friendly Design	(5)
		07. User Comfort	(4)
SOFTWARE	USE & SOCIO-PERCEPTUAL VALUE	08. Diversity & Intensity of Use	(2)
		09. Social Activities	(4)
		10. Identity (Image & Character)	(4)
ORGWARE	OPERATIONAL VALUE	11. Provisions (Amenities, Services, Facilities)	(4)
		12. Safety & Security	(2)
		13. Management & Regulations	(5)

图2 Fig.2 评估系统-城市空间元素、城市价值、属性及评估
Evaluation System – urban space components, urban values, attributes and evaluators

等级标准。为了持续改善住宅公共空间的老龄化硬件性能，城市空间分析工具建议的最重要的标准是：增加普及高等教育的选择；改善连接性（通过提供与周围环境中的主要移动路线的更直接的连接以及在发展过程中建立更好的视线来实现）；更好地使用各种流动手段（尤其是负担得起的公共交通、充足的停车设施、指定的出租车招呼站、自行车车道和停车场的选择和可用性），同时给予行人优先权；更好地接近自然风景；以及保护在发展过程中主要的人行道免受天气条件的破坏。这些措施将会很大程度地鼓励积极养老，并提升其便利度、流动性、用户的舒适度和恢复效果。

使用及社会感知价值可以通过以下措施大大得到提高：在居住区域内及周围提供更多样化的活动（尤其是在指定的运动、娱乐、和社区结合区域）；更好的正式与非正式的座位选择（特别关注座位的多样性，包括单座、团体座、为特定人群提供的带靠背和扶手的座位、幼儿及哺乳期妇女专座，等等）；鼓励加强与空间的交互作用（通过提供灵活的和/或可移动家具及其它交互性元素来实现）；以及增强"形象性"。集体感和空间归属感，然后是居家老龄化可以通过从正式与非正式交互作用设置不同的环境来引发和维持，实现某种程度的对空间的控制（通过其灵活性来实现），并结合显著的设计特点。最后，居住区域中城市空间的操作性价值，包括导向标识、安防及规划，可以通过以下措施进一步提高：提供更好的卫生、环境照明及信息设施；社会和医疗服务；更好的安防系统；及居民积极参与空间管理。

同时还建立了特定标记和描述符的等级标准。例如，针对高架公共空间，如屋顶花园或天桥，最关键的标准是要提供方便不复杂的及清晰的通用进入点和导向标识，可用的环保设施，在天气状况不佳时提供保护，使用和座椅设施的多样性、隐密场所、卫生设施、照明、安全及使用规定及使用时间。识别半开放或非开放空间的重要因素包括：清晰的视线、自行车设施、接近绿化及水文景观、空间内利用的多样性、座椅设施的选择性及灵活性及使用规定。此外，TUSA还提供了在同时选用两个或多个标记时突出更为关键或基础标准的方式。

日本东京东云柯顿街

综合得分最高的城市居住空间就是日本东京的东云柯顿街的大型密集住宅群，于2005年建设完工[图6]。它在公共住房及公共空间中引入了新的概念，将住宅与办公（SOHO）、商业设施和公共娱乐设施以一种不寻常的方式结合在一起。

调查研究内部和外部空间之间的关系是本项目最重要的

图3 Fig.3
环状图-城市空间价值图表（左）及城市空间价值饼图（右）
Circular Charts – Urban Space Value Diagram (left) and Urban Space Value Pie Chart (right) for Old Man Square, Chinatown, Singapore

图4 Fig.4
将世卫组织老年友善属性纳入城市空间框架
WHO's age-friendly attributes translated into the Urban Space Framework

特质之一，因此做出了具体的设计及产生了对半私人空间的集中使用——所谓的"户外生活空间"（OLS）。通过挖空外立面容纳公共区域，利用彩条装饰的可移动木板调整公共区域隐密程度。这样，就保证了公共区域与底层和二层中心公共空间的密切而灵活的交流。S型曲线道路横穿大厦中部，与购物商场和地铁站相连。该街道上儿童设施、公共建筑、商业设施、小型袋状公园林立，还可以看到附近的辰巳运河，让你享受独特的空间体验[图7]。6个线性广场插入道路中，形成OLS的主干。这种由各种公共和半公共空间形成的三维网促进了频繁的活动和社会互动，伴随有各种社区联合项目，如临时的跳蚤市场以及在屋顶花园共同种植蔬菜。各层提供多样化的绿化植物，形成了软屏障，勾画出与众不同的子空间。如何通过动态空间布局，利用公共和半公共空间、底层和高层、城市街区和环境的互动来加强各个年龄段的积极生活和社会联系，东云柯顿就是个例子。这个城市空间具有显著的指示牌、创新的应用灯光和灵活的座位，是一个很好的关于安全而吸引人的环境例子，识别性日益提高和导向标识也日益增多[图8]。

新加坡牛车水广场（老人广场）

尽管硬件和组织件存在一定的局限性，但调查结果强调的是老人广场具有其中一项适合老龄化人口的最佳软件性能[图9]。老人广场是处于佛牙寺和新加坡市中心牛车水大厦之间的一个城市空间。由于该城市空间渗透性高，人们必须步行穿过老人广场进入主要活动区域——活动广场或象棋博弈区。然而活动舞台只会在偶尔的表演或节庆活动中（如元宵节或春节）才活跃起来。每天大部分的活动都在掩蔽区进行。由于热带气候阳光强烈且多雨，掩蔽区很受欢迎，它变成了一个用来社交、下象棋和西洋棋、休息或仅供人们观看的空间。它尤其吸引居住在附近的老年人[图10]。

该广场的用途独特，为区域创造了很强的识别度和强烈的社区意识，对于居家老龄化很有必要。然而混合规划使用的缺乏使它无法吸引更广泛多样的用户。这个空间可以容纳数量有限的固定但多样的家具，如象棋桌和

长凳，因为大多数人只是聚在象棋桌四周。人们也会自己带上便携式椅子。

广场远离主干道，利于步行，尽管没有监视摄像机或守卫，但仍很安全。总体感觉这个广场是一个"自我监察的社区"，用户自己监视，这样增加了用户对空间管理的参与度。然而由于老人们会随地吐痰、乱扔垃圾，所以该场地的维护就比较差。

记住，新加坡的目标是给市中心重新注入活力，该项研究把老人广场看作一个重要的城市元素。它不仅描绘了强烈的地方认同感，还建议用一些方法在市中心为老年人创建新的城市社区空间。

结论

然而对老年人的被动支持（通过住宅单元和周边环境的选择、规模、通用设计以及社区标准节目）似乎成为了已确立的规范。老龄化人口的城市化空间结构强调要活跃的设计措施，支持后代在各种密集和混合型的城市环境中成功老龄化。为了鼓励老年人在不同的年龄阶段增强他们的能力和竞争力，城市空间要充当一个活跃的角色以及支持性的治疗器具，为他们提供挑战而不仅仅是提供方便。运动场——运动和康乐活动空间常常能达到这一点。然而为了充分促进健康的生理和心理老化过程，城市空间设计师需要找到方法，更好的利用多感官体验和加强用户和空间的互动。借助（如自然、棋类游戏或艺术品）交互特性、材质、材料、环境的多样性和颜色，探究强调触感体验和以感官为主导的设计，指示牌对于标记新公共空间的硬件至关重要，这种公共空间对老龄化人口很敏感。此外，不应把通用设计看作是没有文化的。各种研究表明老年人常常聚在特定的场所，并能更好地对具有文化特性的指示牌作出回应。因此具有本土颜色、指示牌和艺术文化特性的设计可以改善导向标识，鼓励体育活动（如使用楼梯而不是电梯），以及增强归属感从而支持居家老龄化。

调查结果表明创建多代空间的关键是重叠活动而不是隔离。重叠活动会刺激不同年代之间的情感依恋、尊重以及同情。许多研究（例如，Jan Gehl[8]）都指出老年人和孩子常常同时去相似的场所，他们通常喜欢聚在那些可以遇见其他人的地方——通常靠近停车场和大楼入口。有些地方没有可供面对面互动的共用空间，如在新加坡的美食街和架空层，这样可能会使老年人的城市空间变得不那么重要。

总之，城市空间分析工具（TUSA）是一个有益的新方法，帮助我们为大部分处在高密度环境中的用户了解、分析、指导规划和设计良好的城市空间。此外，城市空间分析工具的过滤功能使我们能够看清楚相关的特定关键需求和空间属性，如针对老龄化设计的更准确、更敏感的城市空间分析和设计建议。

致谢

本文讨论的研究项目是"高密度可持续环境的城市空间规划"项目中的一个子项目。该项目由新加坡国立大学设计与环境学院亚洲城市可持续发展研究中心与新加坡市区重建局、新加坡国家公园局及新加坡建屋发展局联合开展。我们还要衷心感谢来自新加坡国立大学建筑与设计学院的众多项目协作者的贡献，包括项目初期的首席调查员谢立敏博士以及大卫希·布恩萨姆博士、恩文·威瑞博士、帕特里克·詹森博士和王才强教授。

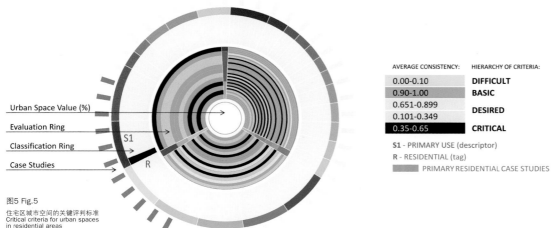

图5 Fig.5
住宅区城市空间的关键评判标准
Critical criteria for urban spaces in residential areas

Urban Space in a High-density Environment

The turn of this century has been characterised by the dramatic rise of the urban population and urban development globally. As of 2007, over half of the world's population lives in cities, and it has been estimated that such a number will increase to two thirds in the next 35 years. It is projected that between 2000 and 2030, Asia's urban population will increase from 1.36 billion to 2.64 billion.

With such urbanisation and population trends, as well as growing quests for environmentally and socially sustainable planning, the ways we understand, conceptualise, design, and utilise urban space is once again challenged. Once densities increase, space becomes a precious commodity charged with intense competition, tensions, and negotiations between dynamic and diverse groups of users. Thus, uncovering new ways to (re)create good urban spaces for all user groups that would fit well in dense urban conditions yet be vibrant and environmentally and socially sustainable would be the main challenge.

Urban Space Framework

Past research has predominantly focused on familiar models of urban space such as squares, plazas, streets, or parks. We still know too little about the increasingly hybrid and dense urban space typologies that are emerging today. The existing literature and research offers no specific values or criteria that are purposefully tailored for understanding and evaluating new urban spaces in high-density and high-intensity environments. With such new conditions, it would be necessary to consider a difference in the nature of the urban spaces rather than a degree of difference from the status quo, which involves speculative dimensions with the main focus on complex relationships between spatial typologies, programmes, and methods of utilisation and management.

In response, a main objective of our research was to discern the relationships between density and quality of urban public spaces based on key criteria that affect a space's performance. This was done by developing the Urban Space Framework. Thereafter, an objective was to derive a system for cataloguing, classifying, evaluating, analysing, and speculating on hybrid urban space typologies, conditions, and performances.

Method

Based on the urban design theory and research of Jan Gehl, Carmona, and Shaftoe among others, an initial conceptual research framework was proposed, and 26 Singaporean and 25 international urban spaces in high-density contexts were documented and analysed. Documentation of case studies involved structured first-person observations, including on-site recording of spatial features and intensities of activities using observational checklists, mapping, drawing, photography, textual description, and occasionally interviewing, as well as a comprehensive review of secondary sources.

Framework

A set of preliminary case study analyses served to refine the research framework, resulting in an original Urban Space Framework. The original framework recognises three concurrent key components that constitute the properties and performances of urban spaces, namely: hardware, software and orgware. Hardware refers to the design values of urban space – its physical and geometrical properties. Software combines activities and socio-perceptual values of space. Orgware refers to operational and management aspects of public space, and often links or overlaps with hardware and software components [Fig. 1].

The Urban Space Framework consists of classification and evaluation systems, which are descriptive mechanisms used to identify critical urban space attributes and assess their performances.

Classification System

The classification system consists of a number of descriptors and tags used to describe and categorise urban spaces into default primary, secondary, and hybrid typologies. Based on a primary descriptor – S1: Primary Use – the system recognises five default urban space typologies, namely urban spaces in residential areas, recreational zones, urban centres, integrated mixed developments, and infrastructural transit-led areas.

Evaluation System

The evaluation system consists of 92 criteria, grouped hierarchically into 46 evaluators, 13 attributes, and five urban values, used both to describe and assess the performance of urban spaces [Fig. 2].

Scoring System

The scoring system allows urban spaces to be assessed based on an evaluation checklist, including all criteria proposed in the evaluation system. A space scores '1' if it meets a criterion. The final score is a sum of all met criteria and forms the overall Urban Space Value, represented in a percentage.

Circular Charts

The scores generated through our research were presented using two circular charts – the Urban Space Value Diagram (which also served as an evaluation check-list) and the Urban Space Value Pie Chart [Fig. 3]. The circular charts are divided into three segments, each representing one urban space component (hardware, software, and orgware). Each segment consists of stripes, and each stripe represents one evaluative criterion. If the criterion is met by the space, the stripe is coloured. While the Urban Space Value Diagram shows an overall performance of a particular space, the Urban Space Value Pie Chart shows the performance of a space for each component.

Analysis

Scores for each criterion for all urban spaces within the same typology were investigated using consistency analysis. A hierarchy of criteria was established, based on average score deviation for each criterion. The hierarchy recognises: (a) basic criteria – criteria most frequently met by all spaces of the same type; (b) difficult criteria – rarely or never met by all spaces within the same typology; (c) desired criteria – inconsistently met by all spaces; and (d) critical criteria – most inconsistently met by all urban spaces of the same type. The focus was on critical criteria, since they are the most feasible to meet and thus have the highest corrective potentials to improve the overall urban space performance.

Synthesis – Tool for Urban Space Analysis (TUSA)

As a synthesis of documentation, classification,

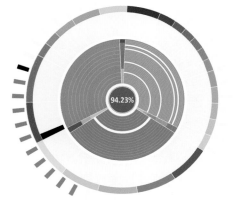 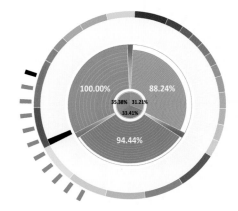

图6 Fig.6 东京东云柯顿街老龄化城市空间价值图表（左）及饼图（右）
Shinonome Codan Court, Tokyo – Ageing Urban Space Value Diagram (left) and Ageing Urban Space Value Pie Chart (right)

图7 Fig.7 东云柯顿街S型街道、绿地及临时社区活动
Shinonome Codan Court, Tokyo – S-shaped street, greenery, and occasional community events (flea market)

图8 Fig.8 东云柯顿街环境照明、显著特征及灵活城市家具
Shinonome Codan Court, Tokyo – ambient lighting, strong identity, and flexible furniture

evaluation, and analysis, an integrated computational tool has been developed – the Tool for Urban Space Analysis (TUSA). Besides being an interactive catalogue, TUSA is primarily used as an analytical and speculative tool, providing fruitful means to study, understand, and guide the urban space design process.

Ageing Population and Urban Space – Demographic Trends and Initiatives

Global demographic predictions state that the number of people aged over 65 will rise from the current 390 million to 800 million by 2025 – reaching ten per cent of the total population. More dramatically, in Asia, the 65-plus age group will increase by 314 per cent – from 207 million in 2000 to 857 million in 2050, reaching 18 per cent of the total Asian population.

Singapore has one of the fastest ageing populations in Asia. According to IMC's Report on the Ageing Population, Singapore's population over the age of 65 will increase from 7.3 per cent in 1999 to 19 per cent by 2030. In other words, by 2030, the current ratio of one elderly person per 12 Singaporean citizens would almost triple to one in five. Moreover, the elderly of the future will be more educated and more independent; they will work longer and will thus have different needs and aspirations than the elderly of today.

"Active Ageing" and "Ageing-in-Place"

Recognising the potentials (rather than solely the limitations) of the ageing population trends, in the late 1990s the World Health Organization (WHO) adopted the term 'active ageing' as part of an initiative to ensure a balanced, healthy, and pleasant ageing process. It is defined as "the process of optimising opportunities for health, participation, and security in order to enhance quality of life as people age". It recognises six groups of key active ageing determinants, namely: social, economic, health and social services, behavioural, personal, and physical environment, with culture and gender as cross-cutting determinants.

Active ageing is now one of the mainstream concepts accepted and applied worldwide, including in Singapore, and involves improving the built environment and acting on changing perceptions towards the elderly, de-stigmatisation, and social integration. The main challenges are thus to encourage active ageing and build up the competency and ability of the elderly to stay independent and healthy by providing a holistic approach to supportive built environments.

Although there is an increasing trend worldwide for older people to live alone, the family has traditionally been the main source of support for the elderly in Asia, and Singapore is not an exception. Approximately 85 per cent of Singaporean elderly people live with at least one child. Thus, along with active ageing, "ageing in place" has also been recognised as the key principle to be applied in public housing in order to enable older people to age in their own homes. Ageing in place is defined as "the ability to live in one's own home and community safely, independently, and comfortably, regardless of age, income, or ability level".

Led by active ageing and ageing-in-place principles, Singapore has developed a holistic concept that combines three main elements, namely: "heartware", "software", and "hardware" as described by the Inter-Ministerial Committee. "Heartware" refers to changing social values into positive attitudes and care towards the ageing process and the elderly. "Software" involves community infrastructure, programmes, and care services for diverse groups within the ageing population. Finally, "hardware" includes various aspects of the built environment. Based on such a strategic framework, major local initiatives include provisions of elder-friendly housing, easily accessible barrier-free environments, affordable healthcare and eldercare, and promotion of active lifestyles and well-being.

Urban Space for the Ageing Population in a High-density Environment

Based on an active-ageing framework and investigations conducted in 33 cities and 22 countries, in 2007 the WHO published its first Global Age-Friendly Cities: A Guide for urban planners and designers, which identified the key physical, social, and services attributes of age-friendly urban settings. Eight key attributes were identified, namely: (1) outdoor spaces and buildings, (2) transportation, (3) housing, (4) social participation, (5) respect and social inclusion, (6) civic participation and employment, (7) communication and information, and (8) community support and health services [Fig. 4].

In order to address the higher demands, specificities, and complexities posed by high-density developments and a growing diverse population, an additional feature has been added to TUSA and the initial version of the Urban Space Framework: a "filter" that enables the researcher to assess the urban space performance in regards to more specific aspects. More precisely, recommendations provided by the WHO's Guide were used to narrow down the criteria proposed in the original framework and develop an upgraded Urban Space Framework for Ageing Population.

In addition to the WHO's various aspects of age-friendly urban development, the upgraded framework retained additional criteria found to be relevant for successful ageing in public spaces in high-density conditions. Some of them refer to: provision of good sightlines within and between space levels (as it contributes to understanding of spatial configuration, and enhances wayfinding and safety); cycling facilities (as they enhance active and healthy lifestyles, as well as car-free development); visual landmarks and focal points of activities (as they contribute to the creation of multi-generation spaces, safety, and space identity); diversity and access to greenery (as it supports recreational uses, enlivens perception, and has restorative effects); variety of shade and sunlight conditions (as it increases comfort and choice); type and diversity of formal and informal seating and interactive elements (as they create points of social and multi-sensory interaction with and within space); and ambient lighting (as it contributes to aesthetics and wayfinding, and enhances extended-hour uses). While adding modifications to better address the elderly population, the new framework applies to all spatial typologies and all user groups, inclusive of children and working adults, bearing in mind that everyone ages.

Application and Findings

All selected local and international public spaces were re-evaluated based on the upgraded framework and analysed using the Tool for Urban Space Analysis (TUSA), as described in previous sections of this paper. The initial findings showed that most of the case studies perform the best for the hardware component, while having the lowest scores for the software. Out of 51 spaces, 17 scored 80 per cent and above for the overall Ageing Urban Space Value. The main objective, however, was to arrive to a set of criteria that are critical for both the process of successful ageing and overall urban space performance for selected classification tags – typologies. While such analysis was conducted for all default urban space typologies, key findings regarding the critical criteria and recommended areas for improvements are here discussed for urban spaces in residential areas and urban centres (squares), with a focus on some of the best international and local examples.

Critical Criteria and Areas for Improvement – Urban Spaces in Primarily Residential Areas

Figure 5 diagrammatically summarises the hierarchy of criteria for primarily residential spaces using consistency analysis. The most critical criteria, suggested by TUSA, for substantially improving the ageing hardware performance of residential public spaces are: increasing choice of universal access; improving connectivity (by providing more direct links to predominant movement routes in the surroundings, as well as establishing better sightlines within the development); providing better access to all means of mobility (especially choice of and accessibility to affordable public transport, sufficient parking facilities, designated taxi stands, cycling lanes, and bike stands) while prioritising pedestrians; providing better accessibility to natural features; and protecting major pedestrian pathways within and around the development from weather conditions. Such actions would considerably encourage active ageing and improve convenience, mobility, users' comfort level, and restorative effects.

Use and socio-perceptual value can be critically improved by: providing a greater diversity of activities within and around the residential area (especially in areas designated for exercise, recreation, and community bonding); better choice of formal and informal seating (with special attention given to diversity of seating, including single seating, group seating, seating for specific demographic groups with back support and armrests, seats for small children and nursing mothers, etc.); encouraging stronger interaction with space (by providing flexible and/or moveable furniture and other interactive elements); and improving 'imageability'. A sense of community and belonging to space, and thus ageing in place, can be initiated and sustained by setting a variety of conditions for formal and informal interaction, allowing a certain level of control over space (through its flexibility), and incorporating memorable design features. Finally, the operational value of urban spaces in residential areas, including wayfinding, security, and programmes, can be further improved by: providing better hygiene, ambient lighting and informational amenities; social and healthcare services; better security; and residents' active involvement in space management.

A hierarchy of criteria was also established for particular tags and descriptors. For example, for elevated public spaces, such as roof gardens or bridges, the most critical criteria relate to easy, non-complicated, and legible universal access points and wayfinding; provision of accessible green features; protection from weather conditions; diversity of uses and seating amenities; privacy; hygiene facilities; lighting; security; and regulations of uses and time of use. Critical attributes identified for semi-open or covered spaces include: clear sightlines; provision of cycling amenities; access to green and water features; diversity of uses within space; choice and flexibility of seating amenities; and use regulations. Additionally, TUSA also provides means to highlight mutually critical or basic criteria, when two or more tags are selected simultaneously.

Shinonome Codan Court, Tokyo, Japan

The residential urban space with the highest overall scores is the large-scale dense residential complex Shinonome Codan Court in Tokyo, Japan, construction of which was completed in 2005 [Fig. 6]. It introduces new concepts to public housing and public space, combining dwellings with offices (SOHO), commercial facilities, and public amenities in unusual ways.

Investigation of the relationships between interior and exterior spaces is one of the most significant qualities of the project, which resulted in specific designs and intense usage of semi-private spaces – the so-called 'outdoor living spaces' (OLS). The facades are hollowed out to accommodate common areas where the level of privacy can be adjusted with moveable timber panels decorated with coloured stripes. In such a way, intense yet flexible communication between common areas and the central public spaces on the ground floor and second level is maintained.

A curved S-shaped path crosses the middle of the complex, connecting a mall and a subway station. Lined with children's facilities, communal buildings, and commercial services, as well as small pocket-like parks and views to the nearby Tatsumi canal, this street offers a distinct spatial experience [Fig. 7]. Six linear plazas that are inserted into the path form the backbone of the OLS. This three-dimensional network of diverse public and semi-public spaces fosters intense activities and social interactions, accompanied by various community bonding programmes, such as occasional flea markets and the communal growing of vegetables in the roof gardens. All levels offer extensive and diverse greenery, which is used to form soft barriers and delineate different sub-spaces.

Shinonome Codan gives an example of how to reinforce active living for all ages and social bonding through dynamic spatial layout, with strong interaction between public and semi-public spaces, ground floor and upper levels, urban block and context. With memorable signage, innovative use of lighting, and flexible seating, this urban space is a good example of an inviting and safe environment, with increased legibility and wayfinding [Fig. 8].

Old Man Square, Singapore

In spite of certain hardware and orgware limitations, findings highlight that Old Man Square has one of the highest software performances for the ageing population [Fig. 9].

Old Man Square is an urban space between the Buddha Tooth Relic Temple and Chinatown Complex in Singapore's downtown. Due to the high permeability of the urban space, people are invited to meander past, through, and into the main activity areas – the event square and chess-playing area. While the event stage only comes alive during occasional performances or festive events (such as the Lantern Festival or Chinese New Year), everyday activities mostly take place within the sheltered area. Due to the harsh sun and rain of the tropical climate, the sheltered area has become very popular – a space for social interaction, Chinese chess and checkers, resting, or simply people watching. It particularly appeals to the elderly men who live in the neighbourhood [Fig. 10].

The specific usage of the square has created a very strong identity for the area and a strong sense of community, which is essential for ageing in place.

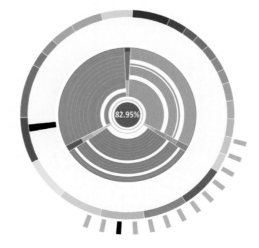 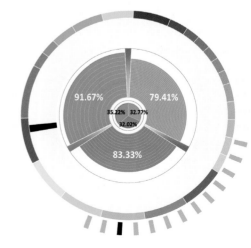

图9 Fig.9　牛车水广场老龄化城市空间价值图表（左）及饼图（右）Old Man Square, Singapore – Ageing Urban Space Value Diagram (left) and Ageing Urban Space Value Pie Chart (right)

图10 Fig.10　牛车水广场有遮挡的棋类娱乐区 Old Man Square, Singapore – sheltered chess-playing area

However, the lack of a greater mix of programmatic usage fails to attract a wider diversity of users. The space provides limited number of fixed yet diverse furniture, such as chess tables and benches, and because of this most of the crowd simply gathers around the chess tables. However, a considerable number of portable chairs have also been brought in by the users themselves. Located away from the main roads, the square is pedestrian friendly and appears safe despite the absence of surveillance cameras or guards. There is a general feeling of 'community self-surveillance', monitored by the users themselves, which increases users' participation in space management. However, the space is relatively poorly maintained due to the fact that the old men sometimes spit and litter on the floor.

Keeping in mind Singapore's goal of re-injecting life back into the city centre, this research sees the Old Man Square as an important urban element. It not only portrays a strong local identity, but also suggests ways of creating new urban community spaces within the city centre for the elderly population.

Discussion and Conclusions

While somewhat passive support for the elderly (through choice, size, and universal design of housing units and immediate surroundings, as well as standard community club repertoire) seems to be an already established norm, the Urban Space Framework for the ageing population highlights more active design measures to support the successful ageing of future generations in various dense and hybrid types of urban environments. In order to encourage the elderly to build up their abilities and competencies at different stages of ageing, urban spaces need to acquire an active role and serve as supportive and therapeutic devices by providing challenges, rather than mere convenience.

This is often achieved with playgrounds – spaces for exercise and active recreation – yet in order to fully stimulate a successful ageing process both physically and mentally, the designers of urban spaces need to find ways to better embrace multi-sensory experience and encourage interaction between space and users. Exploring sensory-charged design with an accent on haptic experience through interactive features (such as nature, board games, or artwork), textures, materials, ambient diversity, colours, and signage is seen as crucial for making the hardware of the new public spaces sensitive to the ageing population. Moreover, universal design should not be looked upon as culture-less. Various studies show that the elderly tend to gather at local-specific places and better respond to culture-specific signage. Thus, culture-specific design, including local colour codes, signage, and art can improve wayfinding, encourage physical activity (such as the use of stairs rather than elevators), increase the sense of belonging, and thus support ageing in place.

Findings suggest that overlapping activities, rather than segregation, can be crucial for the creation of multi-generational spaces; it can incite emotional attachment, respect, and compassion between different generations. It is advocated by a number of studies (for example, by Jan Gehl) that find children and the elderly tend to use similar spaces simultaneously and enjoy gathering and playing at those spaces where they can see other people – usually close to car parks and entrances to buildings. A loss of such common spaces for face-to-face interaction, such as food courts and void decks in the context of Singapore, may diminish this important role of urban spaces for ageing. Interestingly, findings show that the more hybrid urban spaces (in terms of both form and function) generally have a higher ageing urban space value.

In conclusion, the Tool for Urban Space Analysis (TUSA) is a helpful new means of understanding, analysing, and guiding the planning and design of good urban spaces for a wide range of users within high-density conditions. Additionally, with the filter feature, TUSA provides lenses for more specific and sensitive urban space analysis and design recommendations in relation to specific key needs and spatial attributes, such as ageing-friendly design.

Acknowledgements

The research discussed in this paper is part of the research project 'Urban Space Planning for Sustainable High Density Environments' conducted at the Centre for Sustainable Asian Cities (CSAC), School of Design and Environment, National University of Singapore, in collaboration with Singapore's Urban Redevelopment Authority (URA), National Parks Board (NParks), and Housing and Development Board (HDB). The project is funded by the Ministry of National Development (MND), Singapore. We also acknowledge valuable contributions to this work made by Dr Hee Limin as initial principal investigator, Dr Davisi Boontharm, Dr Erwin Viray, Dr Patrick Janssen, and Professor Heng Chye Kiang, School of Design and Environment, at the National University of Singapore, as collaborators for the research.

参考文献 References

1. 联合国人口活动基金会（UNFPA），（2007）《世界人口状况：释放城市增长的潜能》，联合国人口活动基金会。参见http://www.unfpa.org/swop/2007/presskite/pdf/swop2007——eng.pdf [访问于2012年5月21日] UNFPA,State of World Population: People and Possibilities in a World of 7 Billion, New York, UNFPA,2011,available at:http://www.unfpa.org/webdav/site/global/shared/documents/publications/2011/EN-SWOP2011-FINAL.pdf (accessed 21 May 2012)

2. 联合国人口活动基金会（UNFPA），（2007）《世界人口状况：生活在有70亿人口的世界的人及其可能性》，联合国人口活动基金会。参见 http://www.unfpa.org/webdav/site/global/shared/documents/publications/2011/EN-SWOP2011-FINAL.pdf [访问于2012年5月21日] UNFPA, State of World Population: Unleashing the Potential of Urban Growth, UNFPA, New York, 2007,available at: http://www.unfpa.org/webdav/site/global/shared/documents/publications/2007/695_filename_sowp2007_eng.pdf (accessed 21 May 2012)

3. Jan Gehl，（1996）《交往与空间：运用公共空间》哥本哈根：Arkitektens forlag Jan Gehl, Life between Buildings: Using Public Space, Copenhagen, Arkitektens Forlag, 1996

4. Jan Gehl和Lars Gemzoe，（2001）《新城市空间》。哥本哈根：丹麦建筑出版社Jan Gehl and Lars Gemzoe, New City Spaces, Copenhagen, Danish Architectural Press, 2001

5. Jan Gehl，（2010）《人性化的城市》，华盛顿：岛屿出版社Jan Gehl, Cities for People, Washington DC, Island Press, 2010

6. Matthew Carmona等人（2003）《公共场所城市空间：城市设计规模》。牛津建筑出版社 Matthew Carmona et al, Public Places Urban Spaces: The Dimensions of Urban Design, Oxford, Architectural Press, 2003

7. Henry Shaftoe，（2008），《欢乐的城市空间：创建有效的公共场所》。伦敦斯特林梵蒂冈出版社HenryShaftoe,Convivial Urban Spaces: Creating Effective Public Places, London, Sterling, VA: Earthscan, 2008

8. 东西方研究项目中心（2002）《人口与健康研究，亚洲未来人口》。火奴鲁鲁：东西方中心 East-West Center Research Program,Population and Health Studies, The Future of Population in Asia, Honolulu, East-West Center, 2002

9. 部际委员会（IMC）（1999）《人口老龄化报告》。新加坡：部际委员会。参见： http://app1.mcys.goc.sg/portals/O/summary/research/Materials_IMCReport.pdf [访问于2012年5月21日] nter-Ministerial Committee (IMC),Report on the Ageing Population, Singapore, IMC, 1999,available at: http://app1.mcys.gov.sg/portals/0/summary/research/Materials_IMCReport.pdf (accessed 21 May 2012)

10. 世界卫生组织（WHO）（2002）《积极老龄化是一个政策框架》。世界卫生组织 World Health Organization (WHO),Active Ageing: A Policy Framework, Geneva, WHO, 2002, p.12

11. 部际委员会（IMC），《老龄化人口报告》，1999Inter-Ministerial Committee (IMC), Report on the Ageing Population, 1999

12. 疾病防控中心，健康术语"居家养老"，参见http://www.cdc.gov/healthyplaces/terminology.htm[访问于2012年3月7日]Centers for Disease Control and Prevention, 'Ageing in place' in 'Healthy Places Terminology' ,http://www.cdc.gov/healthyplaces/terminology.htm, 2010 (accessed 07 March 2012)

13. 部际委员会（IMC），《老龄化人口报告》，1999Inter-Ministerial Committee (IMC), Report on the Ageing Population, 1999

14. 老龄化问题委员会（CAI），《人口老龄化报告》。新加坡。老龄化问题委员会。参见： http://app.msf.gov.sg/Portals/0/Summary/research/CAI_report.pdf, 2006 (accessed 21 May 2012)Committee on Ageing Issues (CAI),Report on the Ageing Population, Singapore, CAI, available at: http://app.msf.gov.sg/Portals/0/Summary/research/CAI_report.pdf, 2006 (accessed 21 May 2012)

15. 世界卫生组织（WHO）（2007）《关爱老人城市指南》。世界卫生组织World Health Organization (WHO),Global Age-Friendly Cities: A Guide, Geneva, WHO, 2007

16. Jan Gehl，（1996）《交往与空间：运用公共空间》，哥本哈根：Arkitektens forlag; Jan Gehl 和Lars Gemzoe，（2001）《新城市空间》，哥本哈根：丹麦建筑出版社Jan Gehl, Life between Buildings: Using Public Space, 1996 and Jan Gehl and Lars Gemzoe, New City Spaces, 2001

主题 II：
人口老龄化空间和建筑老龄化 3/4
Subject II:
Spaces for the ageing of people, and the ageing of buildings

陈青松
新加坡和深圳雅科本建筑规划公司建筑师和城市规划师

TAN Cheng Siong
Architect and Urban Planner
Archurban Architects Planners, Singapore; Archurban Projects Consultancy (Shenzhen), China

仿生式发展：老龄化的振兴
"Bionicas": Ageing is Rejuvenating

本文为城市发展及相关性的保持提出一种更加生态、人性化与和谐的方式：仿生式发展。有史以来，建筑的寿命始终要大于人的寿命。仿生式发展模式为建筑和社会提供了一起变老并适应不断变化的需求的可能。为了保持人口的"年轻化"，我们首先必须要知道如何保持建筑的"年轻化"。我们必须把建筑看做基础设施，通过改变其内部使用来适应老龄化社会。我们必须发展可以安全、经济、可持续地为不断变化的需求提供服务的建筑行业。这样才能实现城市生活的安全性、创造性与自由度。

This paper proposes a more ecological, more humane, and more harmonious way for cities to grow and remain relevant: 'Bionicas'. Historically, buildings have been constructed to last longer than the human life span. The 'Bionica' model offers a way for buildings and societies to grow old together and adapt to changing needs. In order to keep our population "youthful", we must first know how to keep our buildings constantly "young". We must see our buildings as infrastructure, and adapt their interior uses to the ageing society. We must develop a building industry that services changing needs safely, economically, and sustainably. We will then achieve security, creativity, and freedom in city living.

老龄化是自然规律。不幸的是，老龄化可以使城市的经济发展衰落，引发社会冲突，并导致环境恶化。为了避免城市彻底沉沦，政府当局努力提高人口出生率，延长退休时间，鼓励移民，开展拆迁，并打着城市升级换代的旗号将"新"建筑推倒重建。然而，这种方法却无意中引发了更多的不幸与不安全感。我相信肯定有一种更加生态、人性化与和谐的方式，可以保持城市发展及其相关性。这种方法可以称为"仿生式发展"。

自古以老，建筑的寿命就长于人的寿命。几十年来（第二次世界大战结束以来），人们一直在讨论内在陈旧性理论，灵活持久的概念，以及建筑的代谢本质。我认为，在进行建筑设计时，一定要考虑到建筑可以帮助社会优雅地老去，适应人类不断变化的需求，并以一种和谐的方式适应社会的新陈代谢。

我们现在知道，建筑主要有两大现实：1. 可以存在100年甚至更长时间的通用（普遍存在的）结构；2. 建筑内部的微观功能性部件，可以根据经济与社会的发展命运而迅速变化。城市要建设基础设施，例如道路、桥梁、水道、机场、港口和铁路，这些基础设施的寿命都很长。我们为什么不可以将普遍存在的建筑结构视为基础设施（或者称为"空中土地"），并立法规定应该像我们出租土地一样出租建筑的使用权（整体出租或分层出租）？"仿生式发展"一词指的是，通过不拆除整体建筑而只改变部分建筑空间，可以使城市重新焕发青春。该词来源于仿生系统，意指那些可以充电或者机械化的事物，例如机器。还可以指一种内心状态及意图、态度。我们如何对待城市，又希望城市如何对待我们？仿生式城市应该能够不断发展，让城市不断适应城市居民，让它们不断地发展，就像人要经历成长和死亡一样。

城市与人

有人可能认为城市与人从来没有同步的成长过，从未有过共生关系。城市与人可能不知不觉地互相抵制。人其实没有发言权。城市可能已经存在了数千年，但是从来没有善待过人类。城市一直是战争中的必争之地——征服领土、占有土地、攫取食物、掠取资源，人们住在城市中就做好了承受一切痛苦的准备。

在罗马时代，城市居民不是公民就是奴隶。在封建时代，城市居民是贵族或者佃农。在殖民时代，城市居民是统治者或者被统治者。在工业时代，城市居民是企业主或者是打工者。在单一民族的独立国家，城市居民是政府或者公民。如今，在全球贸易的背景下，城市居民又成为投资者或者工人。尽管城市并未善待过人类——剥削、贫民窟、无家可归，人们仍然对城市趋之若鹜。正因如此，城市仍然处于不断扩张之中。

当今，民选政府与城市公民面临的主要问题是移民以及人口老化。一方面，人们仍在蜂拥进入城市。到2025年，仅中国就将增加3.5亿城市人口，相当于美国的人口总和。人口快速增加和城市化发展将城市变成了大型城市。尽管城市面临土地紧缺，工作岗位不足，人权状况不佳，甚至贫民窟遍布等问题，人们仍然愿意在城市中居住。另一方面，城市居民也在不断反抗，人们不再为这个系统的存续而生儿育女，如今已经出现了逃离城市的迹象。

两次世界大战

在二十世纪发生的两次世界大战中，城市的重要性要远大于人。第一次世界大战是为了争夺商业城市，例如威尼斯，旧上海和伦敦老城；争夺工业城市，例如利物浦，伯明翰和曼彻斯特；争夺军事与政治重地，例如巴黎、柏

图1-4 Fig.1-4　作为租屋升级元素的预制插接阳台
Prefab plug-in "eco terraces" as upgrading elements for HDB blocks

林与莫斯科。而第二次世界大战是为了争夺军事重地、经济发展中心、居住城市和世界贸易中心城市,例如伦敦、柏林、列宁格勒、东京、大阪、纽约、香港、首尔、新加坡、巴黎和阿姆斯特丹。

虽然两次世界大战打着为了人类的旗号,但实际上却是为了争夺权力与控制权。人们被杀戮、迫害甚至被屠杀。城市虽然重要,但是一颗原子弹却可以轻松摧毁一座城市。技术、和平、经济与自由的力量并没有使我们为人类打造更好的城市。但尽管如此,城市仍然在继续扩张。

勇敢的新世界

第二次世界大战结束后,城市发展曾有着光明的未来。重建既是现实需求,也为我们带来了机遇。政治家和全社会都满怀希望。但是事情并非看起来那样简单。相反,却朝着复杂化的方向发展。城市与人的关系变得更混乱、更严重。

当建筑师在1960年代和1970年代开始登上历史舞台的时候,一些具有远见卓识的团体和个人提出了一些令人印象深刻的、以人为本的建议。这些建议具有现代和未来主义的特征,提倡自由、美观和幸福的城市生活方式。尽管看起来充满希望,但是政府当局却对此嗤之以鼻。

克里斯托弗·亚历山大建议全民参与建筑设计。约翰·哈布拉坎提出"开放式规划"的理念;他说,让我们进行开放式规划,设计出普适性的空间——让城市去构建城市的建筑框架,让人民去完善框架内部的一切。建筑电讯派希望城市可以流动地为社区提供教育与文化。新陈代谢派为"土地的有意识性"感到哀伤。菊竹清训认为他设计的"空中住宅"可以使我们搭建空中楼阁,但是人们仍然希望保持与大地的亲密接触。黑川纪章的"居住舱体"无法解决人的全部需求。但是,上述这些观点和想法却与我们如今面临的情况有一些相关性。

1970年代和1980年代,冷战和石油危机是这个世界的重大事件。汽车、郊区的开发和房地产都需要更大的市场与经济扩张。美国与欧洲的需求旺盛,来自亚洲的四小龙国家生产能力惊人。以人为本的乌托邦主义卷土重来。城市成倍扩张,以容纳婴儿潮带来的人口膨胀。来自韩国、台湾、香港和日本的廉价商品源源不断地满足人们的需求。冷战的结束带来了1990年代和2000年代的世界贸易大发展,并使得中国巨龙腾空而起。但是,城市的扩张开始变得无序。金融创新和地产耕作时代到来,有毒资产与金融危机也随之而至。城市开始重点关注资本,将人当做数字来盘剥。但是,城镇化仍然有增无减。令人惊奇的是,在如今这个年代,通过董事会甚至某个个人作出决策,决策者甚至无需见面,只是通过无线的方式开会讨论,数十亿乃至百亿的资金就可以在全世界筹集起来,对整个新城镇进行开发。有了这种投资能力,新城镇被建设起来。新加坡和香港都曾受益于此。房地产正在变成可以交易的商品,吸引着中产阶级和穷人投资地产。当然,我们也看到了银行业的崩溃。城市正在将人当成数字来盘剥。尽管出现了繁荣发展,我们同时也看到了城市的衰落。

再次崩溃

当今,我们面临的问题复杂多样,包括气候变化,金融危机,欧债危机,中东的崛起,中国的计划生育政策,移民,城市化,低生育率及人口老龄化等等。此外,中国的经济增长率可能降低到7%-8%之间;美国的就业情况存在不确定性,金砖五国也在面临着调整。

设计师和开发者并非束手无策:生态城市(例如福斯特与合伙人建筑师事务所设计的马斯达尔城;中国城市规划设计研究院、天津市城市规划设计研究院和新加坡市区重建局规划团队总体规划的天津生态城,以及KPF建筑师事务所总体规划的松岛新城),智能城市(例如由奥雅纳工程顾问公司、索布鲁赫·胡顿建筑师事务所、Experientia实验设计公司设计的锡特拉Sitra Low2No街区,盖里生态都市,以及普兰爱伦谷),以及美丽城市(例如扎哈·哈迪德设计的石头大厦)。明星国际建筑作品提供了一些解决方案的建议。但是,即使是最具时代特征的城市,也仍未能解决人口老龄化带来的全部问题。情况的发展并不同步。城市无法支付退休者养老金,很多城市已经破产,变得不可持续发展。而且这一趋势的迹象愈发明显——压力巨大,不安全性提高,人口迅速老龄化,低生育率等等。有些人开始希望城市变得更加整合、共生、同步。很多人感到有必要建立自己的家庭、拥有自己的住房,老有所依,并参加城市发展的进程。总之一句话,他们需要更大的发言权。但是,城市却无法提供这些。我们需要进行根本性的变革。

未来的仿生式发展

目前,紧凑型的高密度城市正在建造多功能的城市综合体,并从经济规模、便利性、安全性和城市生活的流动性中获益。在新加坡,在墩座上建设起来的超级高层建筑就建在火车站附近,可以提供居住、购物、休闲与娱乐设施。通过内建的升级系统,应该维持这种开发,并使其一直存续下去。

多年来,新加坡一直试图鼓励模块式的、预制建筑的发展,但是收效不是很大。现在,我们奖励那些创新的设计与优秀建筑。同样,我们也必须修订1.建筑法规;2.建筑显现;3.资产价值系统及;4. 城市开发管理,以支持新型城市的创造。

为了保持人口的"年轻化",我们首先要了解如何保持建筑的"年轻化"。目前城市升级换代式的管理方法已经宣告失败,因为这种方法将人与建筑割裂开来看待,而没有看到人与建筑的共生关系。如果我们将建筑看做基础设施;修改建筑的内部的装修和功能,使其适应老龄化社会;并且发展一种可以安全、经济、可持续地服务于不断变化的需求的建筑行业,我们就可以实现城市生活的安全性、创造性与自由度。"空中土地"及基础设施建筑理念可以保持建筑的年轻化,并适应城市人口结构的变化。"仿生式发展"将打造"自我振兴"的城市。

"仿生式发展"是城市发展的未来,建筑与人融为一体,并同时振兴。只考虑房地产、统计数据或者技术是远远不够的。城市还需要考虑到人以及人口的老龄化。我们需要打造城中有人、人中有城的局面。

新加坡的生态层

新加坡通过采取生态化发展框架迎接其面临的出生率下

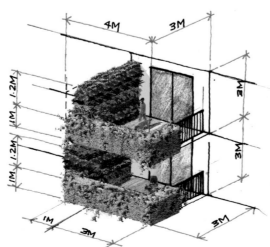
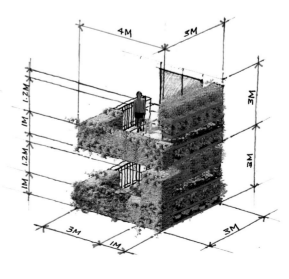

图5 Fig.5　插接阳台轴侧图 Axonometric drawings of plug-in "eco terraces"

降、人口老化、劳动力缩水这三重挑战。独立以后,新加坡的土地成为重要的资源,尤其是对刺激经济发展至关重要。除了几个小块街区外,政府强制收回所有土地。我认为这是引起人们不快的原因之一,因为人们不再对自己的土地有控制权。让我们为新加坡住房发展局所有的公共租赁住房提一些改革建议。

对于新加坡住房发展局现有的公寓:
a) 插入式预制"生态分层",作为升级要素[图1至图5]; b) 有条件永久租赁——业主可以终身拥有空中土地(业主居住的); c) 允许完全的自我填充设计
在新加坡的近代史上,新加坡人只要拥有一套低价的居住单元就感到有了安全感。如今,有条件永久租赁的出现,人们的安全感会更高。政府的工作就是要确保分配的有序性,这样新加坡公民才可以放心的出去工作、娱乐,为自己的健康、财富和心灵而打拼。
生态层振兴计划可以打造新的天际线。此前,新加坡住房发展局曾推出过"主要翻新计划",比起本文提出的生态分层计划,上述计划实施起来要复杂得多。新加坡住房发展局的实体"翻新"对新加坡人来说是实实在在的诱惑——对卧室进行扩展,或者安装新电梯。这些都不是我说的仿生式的发展;只不过在面积上扩大了三四个平方米而已。仿生式发展指的是改变空间,让它与您一起老年化。预制的插入层的想法非常有用——居民可以选择将空调或者洗衣机放到里面,也可以在里面种植蔬菜。新加坡住房发展局的单元非常简单。如果居民年

纪稍大或者生活方式发生变化,生活在这样的单元里很难会感到舒适。插入生态层的成本也不高。我认为全部拆除并且重建并非一个很好的发展方向,但是这种渐进式的变革可以提高人们的安全感。

空中土地——新加坡的未来公寓
我们来想象一下建筑的皮肤。我建议建设一个在地面上建筑的"空中土地"基础设施。'空中土地'[图6—图12]由政府作为基础设施进行开发,就像道路、桥梁、港口、学校、机场和大学一样。政府当局有接入能源、水等市政供应的办法与技术。它们将空中土地分成不同地块,分层销售。居民利用预制插入翻新系统这一新行业供应的预制部件来建设自己的新家。建筑法规要向家有健康老人的繁忙的工作家庭进行倾斜。这样,这些老人可以帮助抚养家里的孩子,同时又有着自己的老年生活。
为了建设新加坡房屋发展局未来的公寓,新加坡将推出:
a) 有条件的永久租赁"空中土地"(分层地契); b) 政府负责规划并建设"空中土地"基础设施; c) 开发一种预制的、插入式的、自己动手的与生长和变老的新陈代谢过程有关的技术; d) 制定一个包括专业人士、建筑商和投资商的系统,实现最低预算,达到最合适的标准; e) 促进规划、设计、建筑管理、融资与体系,鼓励人们在自己的城市居家养老。

新加坡的有条件永久租赁制度可以鼓励产生一种归属感。将现有的99年租赁期改成有条件的永久租赁可以打造一种新的体制,在这个新体制中:
a) 没有"空中土地"的长子或者长女可以继承; b) 如果没有可以继承的子女,则要将"空中土地"归还给政府。政府可以出售,其收益归遗嘱中的继承人; c) "空中土地"的维修保养费用从地产税和物业管理费中支出; d) 社区可以集体提出改进建议; e) 业主可以对内部空间自由翻新或改变,但是要满足安全条件等规定; f) 整体再开发须征得全体居民的同意; g) 无须进行整体销售,因为"空中土地"可以存续一个世纪甚至更长时间。
仿生式城市可以鼓励并欢迎长寿、新生命、社区和社交活动。人是城市的大脑和心脏。建筑是设备与便利推动器,与其中的居民一道,同步进行持续的维护与改进。为了做到最好,新加坡需要有条件的永久租赁制度和仿生式发展。世界上有太多的城市需要振兴。我们可以与他们共享这些经验。通过仿生式城市的发展,我们可以解决诸多问题,例如贫民窟、城市的衰退与人口减少等。

老龄化的振兴
人类热爱城市,城市也需要人类。作为不可分割的一对组合,城市与人需要一起变老、振兴。虽然这两者密不可分,但是城市与人仍然需要学习如何共处。历史上,为了服务于城市,勇士培养了战士,国王使用奴隶,帝国主义国家奴役属国,企业家利用工人,现代的政府则有自己的公民和移民。而如今,公民的选择是不再生儿育女。
正因绝望,才会产生新的观念。各种形状和美学特点的建筑进入设计市场。近年来,各种新颖的观点层出不穷。有些观点可以归入"仿生式发展"的范畴,但是希望这些建筑不是50或者60层楼那么高。有些新颖的甚至是梦幻式的建议包括Atelier Data和MOOV公司的

"向前——阿拉斯加",DCA建筑师事务所的"淡水工厂",以及文森特·卡雷波特建筑师事务所的"蜻蜓翼摩天楼"等。
要想真正将"仿生式城市"变成现实,我们还需要大量的研究和设计工作,否则这一设想将和建筑电讯派、新陈代谢派、克里斯托弗·亚历山大以及约翰·哈布拉坎的很多观点一样,最终烟消云散,成为昙花一现。我们需要的是可以促使人与建筑"共生同步"的城市,关键在于土地,要建设可以进行开放式设计和填充住宅的基础设施。政府不需要盖房子,他们只需要建设"空中土地",并完善有条件的永久租赁(分层地契)。我们应该改革整个的规划、建筑、城市管理与融资体制,使"仿生式城市"蓬勃发展。

Growing old is natural. Unfortunately, ageing leads cities to economic downfall, social conflict, and environmental decay. To prevent the total collapse of cities, authorities desperately try to increase fertility rates, extend the working age, pump up immigration, remove communities, and demolish 'new' buildings in the name of urban renewal and revitalisation. Yet through it, more distress and insecurity unwittingly surfaces. I believe there is a more ecological, more humane, and more harmonious way for cities to grow and remain relevant. This may be called 'Bionicas'.

Historically, buildings have been constructed to last longer than the human life span. Theories of built-in obsolescence, loose-fit long-life, and the metabolic nature of buildings have been discussed for some decades (since World War II). It is my belief that buildings should be designed to allow societies to grow old graciously, be adaptable to people's changing needs, and harmoniously fit society's metabolism.

We now know that there are two realities in a building: 1) the universal (ubiquitous) frame that can last 100 or more years; and 2) the micro functional parts and spaces therein that rapidly change according to the fortunes of economy and society. Cities build infrastructure such as roads, bridges, waterways, airports, harbours, and rail lines that last a long time. May we not treat the ubiquitous frames of buildings as infrastructure (which may be called 'sky land') and legislate their subsequent use on lease as we now lease land (both real or strata)?

The term 'Bionica' refers to a city that can be rejuvenated through a convenient changing of its parts, without demolition of the whole – as in a bionic human. The word comes from bionic systems – things that can be powered or mechanised like machines. It also refers to a state of mind and an attitude of purpose. How do we treat cities, and how do we want cities to treat us? Bionic cities should grow over time. They should allow for alignment with their people. They should allow upgrading, and grow and die as people do.

Cities and People

One could argue that people and cities have never grown together in sync; they have never had a symbiotic relationship. Perhaps unknowingly, they have been working against each other. People have really had no say. While cities have existed for thousands of years, people have not always been treated well by them. Cities have been targeted by war – for territorial conquest, land, food, and resources – and people have been prepared to suffer great pain to live in them.

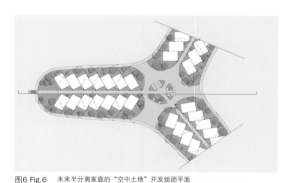
图6 Fig.6 未来半分离家庭的"空中土地"开发组团平面
Cluster plan for "sky land" development of future semi-detached homes

图7 Fig.7 "空中土地"家庭平面 Unit plan for a "sky land" home

图8 Fig.8 "空中土地"开发组团剖面图 Cluster section for a "sky land" development

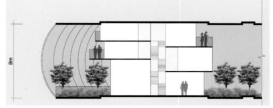
图9 Fig.9 "空中土地"家庭剖面 Unit section for a "sky land" home

图10 Fig.10 未来"空中土地"半分离住宅外观效果图
Rendering of "sky land" development of future semi-detached homes – exterior

In Roman times, the inhabitants of cities were either citizens or slaves. In feudal times, they were nobles or peasants. In colonial times, they were the governors or the governed. In the industrial era, they were industrialists or labour. In nation states, they have become government or citizens. And today, in terms of global trade, they have become investors or workers. Though cities have historically treated people harshly – to the extent of exploitation, slum dwelling, and homelessness – people are hopelessly attracted to them. As such, cities tend to expand.

Today, some of the key issues facing the elected governments and citizens of cities are migration and the ageing population. On the one hand, people keep flocking to cities. By 2025, China alone will add 350 million people – or one USA – into its cities. Rapid population growth and urbanisation mutates cities into megacities. Cities can be short of land, offer people no jobs or rights, and contain slums, yet still people wish to live in them. On the other hand, city dwellers are rebelling – choosing not to procreate to feed the system. Today, there are signs emerging that people are prepared to abandon ship.

Two World Wars

Cities mattered more than people in the twentieth century's two World Wars. World War 1 was fought for the mercantile cities such as Venice, old Shanghai, and old London; the industrial cities such as Liverpool, Birmingham, and Manchester; and the military and political cities such as Paris, Berlin, and Moscow. World War 2 was fought for military cities, economic cities, housing cities, and global world trade cities – London, Berlin, Leningrad, Tokyo, Osaka, New York, Hong Kong, Seoul, Singapore, Paris, and Amsterdam.

The World Wars were fought in the name of the people, but really they were fought for power and dominance. People were killed, persecuted, and massacred. Yet, while cities are so important, they can be so easily destroyed by a single atomic bomb. The power of technology, peace, finance, and freedom has not allowed us to make better cities for people. However, cities continue to expand.

Brave New World

After World War 2, cities held great promise. Reconstruction was both a need and an opportunity. Politicians as well as society in general held great hope. But things were not as simple as they seemed. Rather, they became more complex. Cities and people became more confused, and more severed. When architects held court in the 1960s and '70s, some visionary groups and individuals produced some impressive people-focussed proposals. They were modern and futuristic, promoting freedom, beauty, and a happy urban lifestyle. They seemed hopeful, but authorities were disinterested. Christopher Alexander suggested design by the people. John Habrakan advocated an 'open plan' notion; he said, let's have open plans and universal space – let cities build the frame and the people build up the interior. Archigram wanted cities to walk to deliver education and culture to communities. The metabolists lamented over 'land-mindedness'. Kiyonori Kikutake believed his 'sky house' would free us to build in the sky but people wanted to retain their connection to the land. Kisho Kurokawa could not solve all of people's needs with 'capsule' living. Nevertheless, each of these visions holds some relevance to what we are facing today.

1970s and '80s, the Cold War and the oil crisis took precedence. Automobiles, suburban developments, and real estate needed bigger markets and economic expansion. America and Europe needed so many things, and the four dragons from Asia became very productive. People-oriented utopias took a back seat. Cities multiplied to service the baby boomers. The cheap goods they desired flooded in from South Korea, Taiwan, Hong Kong, and Japan. The end of the Cold War saw world trade boom in the '90s and 2000s. It fed a giant Chinese dragon. Cities, however, suffered from more neglect and more reckless expansion. The era of finance innovations and property farming came into play, along with toxic assets and financial crisis. Cities have been focussing on capital and exploiting their populations as digits. Yet, there is still more urbanisation. It is amazing that nowadays, by the decision of a board or even an individual, through some wireless means, billions of dollars can cross the world and get whole new towns built. New towns are built by these kinds of investing capabilities. Singapore benefited tremendously form that, as did Hong Kong.

Real estate is being turned into a commodity for trading and for enticing the middle class and the poor who want to invest in property. Of course, we now have the banking collapse. Cities are focussing on how to exploit people as digits. At the same time, while booms have been happening, we have also had urban decay.

Breaking Down Again

Climate change, financial crisis, Europe's debts, the Middle East rising, the one-child policy in China, migration, urbanisation, low fertility, and the ageing population are among the many complexities to be faced today. Added to that list are the facts that China's economic growth is expected to slow to 7-to-8 per cent; US job figures are uncertain; and BRICS countries are adjusting.

There are still no lack of ideas and efforts from designers and developers: eco cities (such as Masdar City designed by Foster + Partners and others; Tianjin Eco-City master planned by the China Academy of Urban Planning and Design, the Tianjin Urban Planning and Design Institute, and the Singapore planning team led by the Urban Redevelopment Authority of Singapore; and New Songdo City master planned by Kohn Pedersen Fox), smart cities (such as Sitra-Low2No by Arup, Sauerbruch Hutton Architects, Experientia, and Galley Eco Capital, and PlanIT Valley), and 'pretty' cities (such as Stone Towers designed by Zaha Hadid). International architecture stars propose solutions. However, even age-friendly cities have failed to address the total issue of ageing.

Things are out of sync. Cities are failing to pay retirees. Many are bankrupt. Cities have become unsustainable. Stress, insecurity, rapidly ageing populations, low birth rates – the signs are clear. There is a desire for cities that are part of the people – integrated, symbiotic, and in sync. Many people feel the need to form families and homes, to age in place, and to participate in city development processes – to have a greater say. However, cities cannot deliver. We need something fundamentally different.

Bionicas for the Future

Increasingly, compact, dense cities build multi-function urban complexes to benefit from the economy of scale, convenience, safety, and mobility of urban living. In Singapore, super high-rise towers sit on podiums accommodating commercial, shopping, recreational, and entertainment facilities, which in turn sit over train stations. These developments need to be maintained and sustained through built-in upgrading systems.

Over the years, Singapore has tried to promote, with limited success, prefab, modular and pre-cast construction. We now confer awards to innovative design and construction excellence. In the same vein, we must also revise: 1) building codes, 2) architecture visions, 3) asset value systems, and 4) urban-development management, to support the creation of a new type of city.

In order to keep our population 'youthful', we must first know how to keep our buildings constantly 'young'. The present method of urban management, dealing with renewal and upgrading, has failed and will fail because buildings and people are treated as separate entities, though their relationship is symbiotic. When we finally see our buildings as infrastructure; adapt their interior use to an ageing society; and create a building industry that services changing needs safely, economically and sustainably, we will then have achieved security, creativity, and freedom in city living. 'Sky land' and infrastructural architecture concepts would keep our buildings young and fit to match any demographic change in the city. 'Bionicas' would be self-rejuvenating cities!

'Bionica' is a city for the future in which buildings and people become one, and are capable of being rejuvenated together. It is not enough to consider just real estate, statistics, and technology; cities need to consider people and ageing. We need to make buildings and people increasingly belong together.

图11 Fig.11 "空中土地"开发外观效果图 Rendering of "sky land" development – exterior

Eco Terraces for Singapore

Let's consider how Singapore – a city facing the triple challenge of a declining birth rate, ageing population, and shrinking workforce – could adopt a 'Bionica' framework. After independence, land was an important resource in Singapore – important for encouraging economic growth. The government compulsorily acquired all land, with the exception of a few pockets. This, I feel, may be a reason people do not feel comfortable; they do not have the security of a piece of land. Let's propose some changes for all existing Housing and Development Board (HDB) public housing apartments in Singapore.

For existing HDB apartments: a) Prefab plug-in 'eco terraces' as upgrading elements [Fig. 1 to Fig. 5]; b) Conditional Perpetual Lease (CPL) – land in the sky that a citizen may own for life (owner occupied); c) Permit full self-infill designs. Early in Singapore's modern history, Singaporeans felt secure simply having a low-cost housing unit. Today, citizens would probably feel more secure with a CPL. The government's job would be to ensure such distribution is well managed, and that citizens are free to strike out to work, play, and create health, wealth, and heart.

An 'eco terrace' rejuvenation programme could create a new skyline. Singapore has previously upgraded through the HDB's Main Upgrading Programme (MUP). A simpler process could be implemented for the 'eco terrace' programme. The HDB's typical 'upgrades' are physical enticements for people – a new bedroom extension or a new lift. These are not what I would say could become 'bionic'; they are simply an additional three or four square metres of space. The bionic aspect means you can change the space; it can grow old with you. The idea of prefabricated plug-on eco terraces could be extremely useful – residents could choose to put their air conditioner or washing machine there, or grow vegetables. Singaporean HDB units are rather basic. When residents grow old or their lifestyle changes, it can be difficult to feel comfortable in them. Plug-ons such as the eco terrace would be cheap. I do not believe that total demolition and rebuilding is a good direction, but this kind of progressive change could give people a greater sense of security.

Sky Land – Future Apartments for Singapore

Let's start thinking about the skin of our buildings. I propose a 'sky land' infrastructure of 'landed' properties on decks. 'Sky land' [Fig. 6 to Fig. 12] would be developed by the government as infrastructure – just like roads and bridges, harbours and airports, schools and universities. The authorities have the means and the technology for collecting energy, collecting water, and so on. All they would sell is the decks, which would be subdivided into plots of land. Residents would build their own homes with prefabricated components supplied by a new industry for prefab plug-in renovation systems. Building codes and rules would be geared toward facilitating the busy working family whose old folks are healthy and participating. They could help in raising children while having a life of their own into old age.

For all its future HDB apartments, Singapore could launch the following: a) CPL 'sky land' (strata title); b) Government to plan and build 'sky land' infastructure; c) Develop a prefab, plug-in, DIY technique to relate to the metabolic processes of growth and ageing; d) Set up a system of professionals,

图12 Fig.12 "空中土地" 开放室内效果图 Rendering of "sky land" development – interior

builders, and investors to achieve best economy and best-fit standards; e) Promote planning, design, construction management, financing, and systems to encourage people to age in place with their buildings and cities. A CPL system in Singapore could encourage a feeling of belonging. Conversion of the existing 99-year leases to CPLs could result in a system where: a) First child who has no 'sky land' yet would inherit ;b) Revert to government if no child is entitled to inherit. Government may sell, and proceeds would go to children in will; c) Maintenance of 'sky land' paid by property tax and estate management fees; d) Community may collectively propose improvements; e) Owners may renovate and alter freely internally, subject to safety etc.; f) Enbloc redevelopment would require 100 per cent consent from residents; g) No enbloc sale required as 'sky land' shall last a century or more. Bionicas would welcome long life, new life, communities, and social actions. The people are the brain and the heart of the city. Buildings are the equipment and convenience facilitators that are constantly maintained and improved in sync with the occupants. To achieve the best we can in Singapore, we need CPLs and Bionicas. The world has many cities that need rejuvenating. Experienced gained could be shared with them. Slums, decay, and declining population could all be addressed by Bionicas.

Ageing is Rejuvenating

People love cities; cities need people. They are an inseparable pair that needs to age and rejuvenate together. Though inseparable, cities and people have yet to learn to work together. Originally, to serve their cities, warriors made soldiers; kings made slaves; imperialists made subjects; industrialists made workers; modern governments made citizens and immigrants. Finally, today, citizens choose not to make babies.

Desperation has produced many ideas. Buildings of all shapes and aesthetics flood into the design market. In recent years, there has been a crop of fancy ideas. Some could perhaps be humanised into 'bionicas', but hopefully they are not 50 or 60 storeys high. Some of the fancy, even fantasy proposals have included 'Forwarding Dallas' by Atelier Data + MOOV; 'Freshwater Factory' by Design Crew for Architecture; and 'Dragonfly' by Vincent Callebaut.Much needs to be researched and designed to make 'bionicas' work or the ideas will fade like so many others from Archigram, the Metabolists, Christopher Alexander, and John Habraken. What we need are cities that promote 'symbiotic sync' between people and buildings. The key is land; the construction is infrastructure for open-plan and infill homes. Governments need not build houses. They need only build 'sky land' complete with 'conditional perpetual lease' (strata title). We should revamp the whole planning, building, urban management, and financing system to make 'Bionicas' flourish!

参考文献 References

1. Alusi, A., Eccles, R. G., Edmondson, A. C., 及 Zuzul, T., 可持续发展的城市：逆输还是塑造未来? 哈佛商学院工作论文, 编号11–062, http://www.hbs.edu, 2010 (2012年7月查阅) Alusi, A., Eccles, R. G., Edmondson, A. C., and Zuzul, T., 'Sustainable Cities: Oxymoron or the Shape of the Future?', Harvard Business School Working Paper, No. 11–062, http://www.hbs.edu, 2010 (accessed July 2012)
2. Broto, E., 高密度: 未来之环境, 巴塞罗那, 国际链接出版社, 2011. Broto, E., High Density: Environments for the Future, Barcelona, Links International, 2011
3. Cuperus, Y. 开放建筑初探. Cuperus, Y. 灵myne活建筑, Delft, OBOM, 戴尔福理工大学, 2001. Cuperus, Y., 'An Introduction to Open Building' in Cuperus, Y. Agile Architecture, Delft, OBOM, Delft University of Technology, 2001
4. Habraken, J., 无法回避的问题: 建筑长期趋势及其对建筑教育的影响 in Spiridinodines, C. and Vovatzaki, M. (eds), 塑造欧洲高等建筑教育领域, 建筑教育交易18, 鲁文建筑教育学院, 2003 Habraken, J., 'Questions that Will Not Go Away: Some Remarks on Long-Term Trends in Architecture and their Impact on Architectural Education' in Spiridinodines, C. and Vovatzaki, M. (eds), Shaping the European Higher Architecture Education Area, Transactions in Architectural Education No. 18, Leuven, K.U. Leuven Dept. of Architecture, 2003
5. Hamdi, N., '大众住房, John Habraken 及 SAR (1960–2000), (评论), 建筑评论, 2001年5月1日 http://www.highbeam.com/doc/1G1-75960556.html 2001 (2012年7月查阅). Hamdi, N., 'Housing for The Millions, John Habraken and SAR (1960–2000). (Review)', The Architectural Review, May 1, 2001, http://www.highbeam.com/doc/1G1-75960556.html 2001 (accessed July 2012)
6. Harrison, J. D.,为下一个千年脆弱的新加坡老年人提供住房, 新加坡医学学刊, 1997年第10期, 总第38期, 415-417页 Harrison, J. D., 'Housing Singapore's Frail Elderly in the Next Millennium', Singapore Medical Journal, vol. 38, issue 10, 1997, pp. 415-417
7. Khaw, B. W., 集中购买热降温, 星期日泰晤士报, 2012年1月22日. Khaw, B. W., 'Collective Sale Fever Cooling', The Sunday Times, 22 Jan 2012
8. Koolhaas, R. 与 Obrist, H. U., 日本项目: 与新代谢派对话…, Kohn, Taschen, 2011. Koolhaas, R. and Obrist, H. U., Project Japan: Metabolism Talks…, Kohn, Taschen, 2011
9. Labus, A., 21世纪人口老龄化社会的城市更新概念–波兰城市案例研究, http://www.corp.at, 2012 (2012年7月查阅) Labus, A., 'Concepts of Urban Renewal in an Aging Society in the XXI Century – Case . Studies in Polish Cities', http://www.corp.at, 2012 (accessed July 2012)
10. Ladouce, N., 高密度紧凑型城市, http://ezinearticles.com, 2011 (2012年7月查阅) Ladouce, D., 'High Density Compact Cities', http://ezinearticles.com, 2011 (accessed July 2012)
11. Lane, D., 全球老龄化问题的创新建筑解决方案, 香港房屋协会成立六十周年庆祝大会, http://www.hkhs.com, 2008 (2012年7月查阅) Lane, D., 'Innovative Building Solutions for a Global Aging Problem', Hong Kong Housing Society 60th Anniversary Conference, http://www.hkhs.com, 2008 (accessed July 2012)
12. Lien, L., '婴儿与老龄化: 新加坡已做好应对准备, 新加坡海峡时报, 2012年6月2日 Lien, L., 'Babies & Ageing: Singapore Well-Equipped to Cope', The Straits Times, 2 June 2012
13. Lucchesi, J., 建筑是结构, 不是图片…, http://aiacc.org, 2012 (2012年7月查阅) Lucchesi, J., 'Architecture as the Frame not the Picture … Really', http://aiacc.org, 2012 (accessed July 2012)
14. Nerenberg, J.建筑能否使老年生活更加优雅? http://www.seniorsworldchronicle.com, 2010 (2012年查阅) Nerenberg, J. 'Can Architecture Help the Elderly Age Gracefully?', http://www.seniorsworldchronicle.com, 2010 (accessed 2012)
15. Phua, M. P., 我们还有更好的家庭吗, 新加坡海峡时报, 2012年6月23日. Phua, M. P., 'We Want a Greater Sense of Home', The Straits Times, 23 June 2012
16. 公共空间项目, 克里斯托弗·亚历山大, http://www.pps.org/reference/calexander/, 2009 (2012年7月查阅) Project for Public Spaces, 'Christopher Alexander', http://www.pps.org/reference/calexander/, 2009 (accessed July 2012)
17. Richardson, M. 中国老龄化社会威胁经济繁荣, 新加坡海峡时报, 2012年6月11日. Richardson, M. 'Ageing China Jeopardises Prosperity', The Straits Times, 11 June 2012
18. Sanders, J., 建筑电报派: 未来设计, 国际艺术论坛, 1998年第2期, 总第37期. Sanders, J., 'Archigram: Designs on the Future', Artforum International, vol. 37, no. 2, 1998
19. 新加坡政府, 新加坡社会发展, 青年与体育部, 为解决城市人口老龄化而推出的 '适合各年龄段人口的城市计划' (新加坡社会发展、青年与体育部媒体发布稿, 2011年3月21日), http://www.news.gov.sg, 2011 (2012年查阅) Singapore Government, Ministry of Community Development, Youth and Sports, 'City for All Ages Project to Develop Urban Solutions for Ageing' (MCYS Media Release, 02 March 2011), http://www.news.gov.sg, 2011 (accessed 2012)
20. 新加坡政府, 国民人口与人才分化, 总理办公室, 不定期文章: 公民人口介绍, http://www.nptd.gov.sg, 2012 (2012年查阅). Singapore Government, National Population and Talent Division, Prime Minister's Office, 'Occasional Paper: Citizen Population Scenarios', http://www.nptd.gov.sg, 2012 (accessed 2012)
21. Wächter, S. (Waine, O. 翻译), 数字城市未来之挑战, http://www.metropolitiques.eu, 2012 (2012年7月查阅) Wächter, S. (translated by Waine, O.), 'The Digital City: Challenges for the Future', http://www.metropolitiques.eu, 2012 (accessed July 2012)
22. 华ງ森 中国的空中城市, urbanchinainitiative.org, 2011 (2012年7月查阅) Woetzel, Jonathan 'China's Cities in the Sky', http://www.urbanchinainitiative.org, 2011 (accessed July 2012)
23. 世界卫生组织, 老年友好型全球城市指南: 日内瓦. 世界卫生组织, 2007. World Health Organisation, Global Age-Friendly Cities: A Guide, Geneva, WHO, 2007

Ageing Society and Urban Regeneration Challenges in Hong Kong – Case Studies

主题 II：
人口老龄化空间和建筑
老龄化4/4

Subject II:
Spaces for the ageing of people, and the ageing of buildings

林云峰
香港中文大学建筑学院教授

Bernard V. LIM JP
Professor, School of Architecture
Chinese University of Hong Kong
Hong Kong, China

面对"每个人都会变老"和"城市在变老"的挑战，我们应该广泛讨论并重新审视建筑和城市设计。本文对香港的一些案例进行研究，说明如何通过创新设计解决可持续性的挑战。海坛街的综合开发采用了一种社会反应设计，形成连接破碎节点的公共枢纽。美荷楼的改造保留了怀旧和可持续的生活方式。衙前围村再开发则使用"立体城市"方法设计和创造保护公园兼新开发来保留历史元素。

In relation to the challenges suggested by the statements "Everyone Ages" and "Our City Ages", architecture and urban design should be extensively debated and re-examined. This paper examines some case studies in Hong Kong that demonstrate how the challenges of sustainability can be tackled through innovative design. A comprehensive development in Hai Tan Street adopts a socially responsive design to form a communal hub that may reconnect the fragmented nodes. The revitalisation of Mei Ho House preserves a nostalgic and sustainable living style. The Nga Tsin Wai Village redevelopment keeps the historic elements by designing and creating a conservation park-cum-new development using a "vertical city" approach.

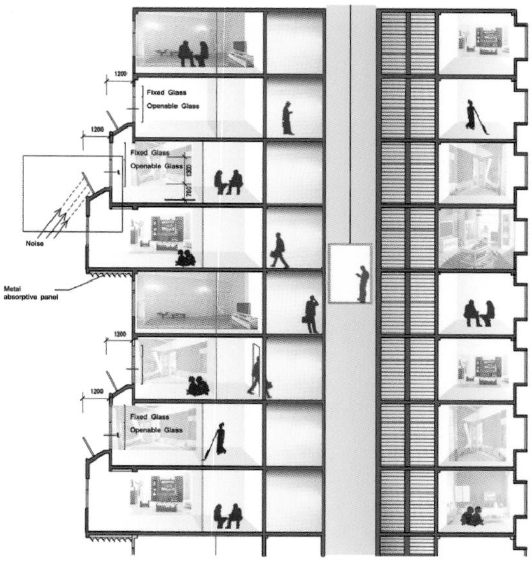

图1 Fig.1　剖面显示交错形式及噪音消减
Section showing staggered form and noise attenuation

人口规模和人口年龄结构

香港人口从1981年的510万增长到2011年的700万。香港总生育率近几十年来从1975年的每1,000名妇女生2,666名小孩降至2011年的955名。结果15岁以下人口比例下降严重。1975年，31%的人口在15岁以下。2011年则只有11.8%。

医疗和保健服务的改善提高了预期寿命，使更多的人能够活过65岁。1975年，新生男婴的预期寿命为70.1岁，新生女婴的预期寿命为76.8岁；2011年这两个数字分别增加到80.5岁和86.7岁，预计2039年将增加到83.7岁和90.1岁。由于出生预期寿命的持续增加，65岁及以上的人口从1975年的5%增加到2009年的13%。预计该比例2039年将进一步增加至28%。人口老龄化急剧上升的背后有两个原因。首先，中国大陆许多年轻人在20世纪70年代后期来到香港。这些人到21世纪10年代中期年龄将会达到65岁。其次，20世纪50年代至60年代香港出现婴儿潮。这些人将从21世纪10年代中期开始加入到"老龄化"人口（即65岁及以上人群）中。

香港城市衰退

城市衰退是现代世界几乎所有地方城市生命面临的最尖锐问题的根本原因。香港也不例外。尽管我们拥有持续的经济成功、商业智慧和高产的劳动力，城市衰退仍然是我们今天在香港面临的最迫切的问题。香港有超过110 000个家庭仍生活在不合规格的房屋中。

贫民窟对香港的生命和财产安全具有不可忽视的威胁。110 000个家庭别无选择地居住在贫民区，而肮脏、衰败的环境和便利设施的缺乏使生活更加艰苦，而香港的其他居民却将其视为理所当然。这些是贫民窟所引起的个人困难。

到2030年，建成超过50年或以上的建筑物数量将达到16 000栋。更可怕的是，多年以来其中很多建筑物都面临着管理不善和缺乏维护的问题。随着破损越来越严重，维护成本增加，将导致不再进行维护。

令人担忧的是超过2 000个不合规格建筑具有非常糟糕的住宿标准。典型的情况是3个或3个以上的家庭共同居住在大约500平方英尺的空间内。
事实是香港将继续其城市衰败之路,生活条件将更加恶化。最近,为了增加租金收入,房东开始将这些公寓划分为更加狭小的住宅单元。这些小隔间被称为"棺材屋"——当地媒体创造出的专门用语。这些生活条件使居民们感到比住在臭名昭著的"笼屋"更加痛苦。

案例研究——香港深水埗市区重建局负责的海坛街住宅开发

深水埗是香港高度发达、人口稠密的地区之一。海坛街住宅开发地显示出若干限制因素:噪音、限高和空气流动。场地位于西九龙走廊南部六车道桥梁结构附近。规划部声明的最大建筑高度为120mPD(超过基准面的米数)。还需要考虑保持邻近空气流动;建筑体积不得过大或对周围通风产生不利影响。

场地也具有若干优势条件:社区、房屋结构和日光照明。该地区位于历史和文化丰富的战略位置,能够激奋社区力量。由于周围古建筑结构多样、典型(现有建筑结构),新的建筑设计能够重新解释廉租房结构和规模,以供人们追忆老社区。南方提供了充足的日光和开阔的视野,但同样也带来了噪音。如果噪音问题能够解决,则可享受充足的日照和开阔视野。

设计理念

"谨慎设计;尊重当地遗产;接轨未来"

为了使深水埗旧城区有一个更好的生活环境,设计充分考虑了环境反应、可持续性和社会回应问题。保护居住单元免受邻近西九龙地区的交通噪音影响,以便享受南方日光和开阔的视野。该地区人口稠密,进行可持续设计时应保证充足的通风。为了保护深水埗当地文化和历史,设计考虑了新开发建筑高度、建筑结构的视觉冲击和对周围现存建筑的影响。同时也考虑了首层规模和矮墙外观的视觉协调。文化和社区也纳入了考虑范围中。

幸福、舒适的建筑——香港城区的生态人居环境

高层建筑采用了错落的建筑形式,不仅是为了协调本地环境,而且也将其作为一种噪音缓解措施,这极大地减小了来自西九龙走廊地区的噪音干扰[图1]。除了对高层建筑采用错落设计以外,还建造了一个5层的非噪音敏感型室内俱乐部大厦作为噪音屏障,以增强噪音衰减效果。普遍的噪音处理方法则是将窗户设计在背离交通的方向,而不考虑视野和采光[图2]。

普遍的综合发展方式是利用整个地区的矮墙设计,建造一栋购物商场以鼓励零售业的发展。引进一条"内街道",连接桂林街和拟建的北河街公共空地,作为一个风道。俱乐部大厦同样采用了错落的设计形式,以减小矮墙体积,增加自然采光,保证"内街道"更好的通风。为了让居民更好的生活,将在公共空地地区、美化屋顶、花坛和个人单元进行种植和垂直绿化。

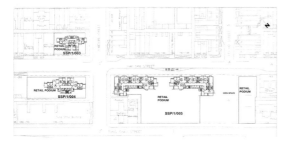

图2 Fig.2 海坛街住宅开发平面图-单边策略
Site plan for Hai Tan Street residential development – single aspect approach

整合社区

场地位于历史和文化丰富的战略要地,其发展旨在恢复旧城区的活力。街区突出的交错排列形式[图3]源自于本地旧建筑的缩退形式和大部分出租楼中非法结构不受控制的增加[图4]。"内街道"增强了步行网络、优化了通行性,因此促进了深水埗地区居民间的交流[图5]。有着传统装饰的商店、餐馆和休闲设施的回廊使人回想起深水埗旧区。

宽敞的空地和分离的建筑使视野障碍最小化,避免了目前和未来发展的墙壁效应。研究表明拟建项目的视觉效果不会破坏或影响山脊线的可视性。

设计被呈交给各利益相关人,包括深水埗区域议会,真实地反应了公众对开发的担忧和看法。根据政府/机构和社区(G/IC)设计的需要,进行了实地考察并与负责人(社会福利部门)进行商议,以了解最终用户的需求。

环境适应和环境保护设计

设计中利用了环保措施可实现可持续性,使产生的废物和需要的资源最小化。设计主要集中于通风和减少噪音干扰,旨在鼓励居民开窗户代替开空调,实现减少耗电的最终目标,并在露天走廊安装运动检测器(红外线型)。夜间关闭所有灯光节约能源。规定在所有公寓和公共厕所都安装双冲洗水箱。自动扶梯主要安装于以商业和娱乐活动为主的花坛以下的位置。当自动扶梯处于预设闲置期间时,会自动减速。

资源循环——使用可再生材料

香港每天都要产生大量的固体废物。根据规定,"生态铺路砖"含有一定比例的回收材料。

功能设计

通过环境评估,以支持有科学的交错设计和内街道设计。交通噪音评估的目的是调查拟重建建筑周围300米内来自各主要道路的噪音情况。评估结果表明,项目总符合率为86%,符合香港噪音限制规划标准与指南总建筑楼数量的80%的标准。进行空气流通性(AVA)评估,以评价拟行人通风环境开发的影响并确定对周围建筑和地势的影响。同时对临近公路网(拟重建建筑周围500米内)车辆尾气排放而造成的空气质量影响进行评估。

结论

结合地方特色和文化环境,内街道的交错设计解决了噪音问题,提供了良好的采光和通风条件。该项目表明,深水埗地区新型住宅和商业发展以及机构设施的设计标准很高,未改变开发商的开发权。

案例研究——香港青年旅社协会城市青年旅馆美荷楼的再生

背景

该项目旨在通过根据香港发展局文化遗产办公室行政长官提出的"协作活化古建筑"的要求,重新利用古建筑,将石硖尾的美荷楼改装成为一栋青年旅舍[图6]。

作为建成的第一栋I类标志性公共住房建筑,石硖尾在香港公共住房历史中有着重要的作用。该项目同时保留了物质文化遗产(建筑文物)和非物质文化遗产(建筑的建造方法);确保了操作是可行可持续的;旨在通过保留特色定义元素带来美荷楼的价值。

设计

美荷楼活化项目的建筑设备设计和安装考虑了作为城市青年旅舍操作和维护的方便性[图7a和图7b]。整体的建

图3 Fig.3 设计方案中的交错形式 The staggered form of the design

筑设计方法涉及:

• 环境舒适性和可持续性设计。最大化自然光照,同时最大化旅社内的被动换气,以获得一个健康的生活空间和良好的室内空气质量。

• 共享空间。流通和大厅空间设计应互相平衡,且应种植绿色植物。设置"公共空间"、"互动区域"和"公共休息室",以满足邻居和公众活动的需要。

• 多层旅社的特殊要求。设计将垂直循环与开放式走廊和共享空间融合,以供互动和使用。

反映生活历史的住房博物馆

保留了基本居住单元以及修复外观的住房博物馆使公众能够重新找回集体回忆[图8a和图8b]。开展描述过去生活方式和社会转型的专题展览。形成社区历史地标并加强公众互动。采用悬浮软性显示器板系统,以缓解改变展览的要求,以及防止原墙面的损坏。

舒适的客房

以三星级酒店为目标,客房应提供一个舒适安逸的环境[图9]。一般青年旅社协会的常客都来自世界各地,他们将会享受到宾至如归的感觉。

案例研究——衙前围村保护项目

"丢失"的历史

衙前围村属于城市地区中的一个围区[图10]。采用的保护方法表明在现存的条件下,不值得重建该村庄,而应建造一个保护公园。设计将采用"设计保护"的创新概念,以保留村庄的三大遗址,即:村庄警卫室、嵌入式碑碣和天后庙以及连接三大遗址的中心轴。为了突出村庄的600年历史,这些被设置为保护公园的核心设计元素,同时可平行开展住宅重建。

案例研究——总结

项目说明了以下价值:

• 连接性。互联互通的街道格状网使得交通分散,步行便利。高质量的步行网络和公共领域使得步行变得非常舒适。

• 多功能设计和多样性。该地区应有商店、办公楼、公寓和住宅,使不同年龄、不同收入、不同文化和不同种族的人能够自由生活。

• 高质量的建筑和城市设计。强调漂亮、美观、舒适以及地方感。人文尺度建筑和幽美的环境滋养了人类精神。

• 传统的社区结构。将公共空地设计为城市艺术。将老年人传统的社会网络考虑到设计当中。

• 可持续性。通过利用环境友好型科技和尊重自然系统,将开发和经营对环境的影响控制在最低水平。

图4 Fig.4 区内老建筑 Old buildings in the district

Population Size and Age Structure of the Population

The population of Hong Kong has grown from 5.1 million in 1981 to seven million in 2011.The total fertility rate in Hong Kong decreased in recent decades from 2,666 births per 1,000 women in 1975 to 955 in 2011. As a result, the proportion of persons under 15 years of age is decreasing. In 1975, 31 per cent of the population was aged below 15 years. In 2011, the corresponding figure was 11.8 per cent.

Life expectancy has increased as a result of improved medical and health services, enabling more people to live beyond the age of 65. In 1975, a newborn baby boy was expected to live 70.1 years and a newborn baby girl to live 76.8 years; these figures increased to 80.5 and 86.7 respectively in 2011 and are projected to increase further to 83.7 and 90.1 in 2039. Owing to the continued increase in life expectancy at birth, the proportion of those aged 65 and over has increased from five per cent in 1975 to 13 per cent in 2009. The proportion is projected to increase further to 28 per cent in 2039.

There are two reasons behind this sharp increase. Firstly, many young people from mainland China came to Hong Kong in the late 1970s. These people will be 65 years old in the mid-2010s. Secondly, a baby boom occurred in Hong Kong in the 1950s-to-'60s. These people will join the "old-age" group (i.e. those aged 65 and over) from the mid-2010s.

Urban Decay in Hong Kong

Urban decay is a root cause of the most acute problems of city life almost everywhere in the modern world. Hong Kong is no exception. Despite our enduring economic success, our commercial resourcefulness, and our highly prolific labour force, urban decay remains one of the most pressing issues we face in Hong Kong today. More than 110,000 families in our community still live in homes that are substandard.

Slums create threats to lives and property in Hong Kong on a scale that cannot be ignored. Daily life for the 110,000 families who have no choice but to live in slums is made wretched by the filth, decay, and lack of amenities that the rest of the community takes for granted. These are the personal difficulties that slums cause.

By 2030, the total number of buildings aged 50 years or more will reach 16,000. What makes the situation even more dreadful is that many of them have suffered from poor management and minimal repairs over the years. As dilapidation gets worse, the cost of

图5 Fig.5 内部街道 Internal street

图6 Fig.6 石尾 的荷美楼改装为青年旅社
Design for the re-use of Mei Ho House at Shek Kip Mei Estate as a youth hostel

maintenance increases, and therefore it does not take place.

The alarming state of over 2,000 substandard structures is matched by the awful standards of accommodation that they provide. Typically, three or more families are packed into the average living unit of around 500 square feet.

The fact is that forms of urban decay will continue to emerge in Hong Kong, and living conditions will become even more degraded. In the hunt for increased rental yields from these buildings, owners have recently begun to subdivide the flats into even smaller units. These cubicles are now known as "coffin homes"– a term coined by the local media. In these conditions, the dwellers are even more miserable than those in the already notorious 'cage homes'.

Case Study – Hai Tan Street Residential Development for the Urban Renewal Authority, Sham Shui Po, Hong Kong

Sham Shui Po is one of the first developed and most densely populated districts in Hong Kong. The site of the Hai Tan Street Residential Development presents several constraints: noise, height limit, and air flow. The site is situated next to a six-lane bridge structure, the West Kowloon Corridor, at the south; this presents a noise problem. In terms of height, a maximum building height limit of 120mPD (meters above principle datum) is declared by the Planning Department. The preservation of air flow around the neighbourhood also requires consideration; the building bulk should not be too massive or adversely affect air flow.

The site also offers several opportunities: community, building fabric, and daylighting. The site ispositioned at a strategic location in a context of rich heritage and culture, which encourages community strength. With the rich and characteristic fabric of surrounding old buildings (existing building fabric), the new design could reinterpret the tenement house fabric and scale to reminisce about the old community. The south offersdaylight and an open view, but also noise.If the latter could be solved, the former could be explored.

Design Concepts

"Designing with care; respecting the local heritage; integrating with the future."

To achieve a better living environment in the old district of Sham Shui Po, the design addresses issues of environmental responsiveness, sustainability, and socialresponsiveness.The residential units shall be protected from the traffic noise of the adjacent West Kowloon Corridor to enjoy to southern sun exposure and the open view. A sustainable design approach balanceswith the requirement of ensuring adequate air ventilation in such a dense urban area.To preserve the local culture and heritage of Sham Shui Po, the design considers the visual impact of the new development in terms of height, building fabric, and impact on the existing surrounding buildings.It also

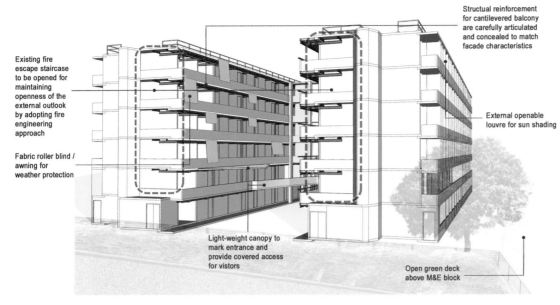

图7a Fig.7a 针对新建筑功能的设计方法及设施配套
The design approach and installation of services addresses the new building use

considers visual harmony at ground level in terms of scale and the appearance of the podium. Culture and community are taken into account.

Architecture of Well-being and Adaptation – Ecological Living in Urban Hong Kong

The residential towers adopt a staggered building form, not only to reminisce on the local ambience, but also to act as a noise mitigation measure, which significantly reduces the heavy noise disturbance from the West Kowloon Corridor [Fig. 1].Aside from the staggered design of the residential tower, a five-storey indoor clubhouse block, which is not noise sensitive, serves as a noise barrier to enhance the noise attenuation. The prevalent noise strategy employs a design in which windows face the direction without traffic irrespective of view and daylight exposure [Fig. 2].

The prevalent composite development employs a design with a podium occupying the entire site,with a shopping mall to encourage retail activities. An "internal street" is introduced to connect Kweilin Street to the proposed public open space at the closed section of Pei Ho Street. It servesas a ventilation passage. The clubhouse block also adopts a staggered design to reduce the bulk of the podium, enhance natural lighting, and allow better air ventilation for the "internal street". For the enjoyment of residents,planting and vertical greenery are proposed for the public open space, landscaped roof, podium gardens, and individual units.

Integrating with the Community

The site has a strategic location in a rich heritage and cultural context, and the development aims to revitalise the old community. The proposed

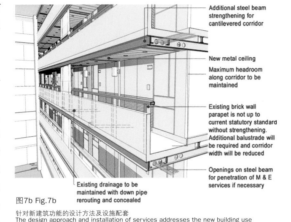

图7b Fig.7b 针对新建筑功能的设计方法及设施配套
The design approach and installation of services addresses the new building use

staggered form with protrusion blocks [Fig. 3] was derived from the setback features in the old local buildings and the uncontrolled proliferation of illegal structures in most tenement buildings [Fig. 4].The "internal street" enhances the pedestrian network, optimises accessibility, and thus promotes neighbourly interaction between residents in Sham Shui Po [Fig. 5]. A cloister with traditionally decorated shops, restaurants, and leisure facilities on two sides recallsthe old Sham Shui Po district.

An ample area of open space and the separation of the towers minimises visual obstruction and avoids the wall effect for both the existing and future developments. The visual impact of the proposed development was studied to demonstrate that it would not breach or affect views of the ridgelines of mountains.

The design was presented to various stakeholders including the Sham Shui Po District Council, which truly reflected the public's concerns and views towards the development. As required for the design of a government/ institutional and community (G/IC) facility, site visits and consultation with the operator (the Social Welfare Department)were carried out to understand the needs of the end users.

Design with Environmental Adaptation and Care

Sustainability can be achieved through the use of environmentally friendly measures in the design to minimise the waste generated and resources needed.The design focuses on ventilation and the reduction of noise disturbance.It aims at encouraging residents to open windows instead of turning on the airconditioner, and achieve the ultimate goal of reducing electricity consumption.Motion detectors (infrared type) are specified forthe Sky Corridor.At night, all lighting is turned off during unoccupied periods to save energy. Dual-flush water tanksare specified for all flats and public washrooms.Escalators will serve the levels below the podium gardens, where commercial and recreational activities dominate. When the escalator is idlefor a pre-determined period, it will slow down.

Resource Loops – Use of Recycled Material

In Hong Kong, a huge quantity of solid waste is produced every day. "Eco-paving blocks" are specified, which contain a proportion of recycled materials.

Functionalistic Design

An environmental assessment has been carried out to support the staggered design and internal street on scientific grounds. The assessment was carried out to survey noise from all major roads within 300metres of the proposed redevelopment. The assessment results indicate that the total compliance rate for the project is 86 per cent, which fulfils the requirement of not less than 80 per cent of the total number of flats complying with the Hong Kong Planning Standards and Guidelines noise limit.An air ventilation assessment (AVA) was conducted to assess the impacts of the proposed development on the pedestrian wind environment and determine the effects of the surrounding buildings and topography.The impact on air quality caused by the vehicular emissions from the adjacent road network (within 500metres of the proposed redevelopment)was also assessed.

Conclusion

Incorporating local characteristics and the cultural context, the staggered design with its internal street solves the noise issue, and offers good daylighting and ventilation. The project reveals a high design standard for Shum Shui Po with new residential and commercial development, as well as institutional facilities, without scarifying the development rightsof the developer.

Case Study – Revitalisation of Mei Ho House as City Hostel for Youth Hostels Association, Hong Kong

Background

This project aims to revitalise Mei Ho House at Shek Kip Mei Estate and transform it into a youth hostel [Fig. 6] through sensitive adaptive re-use of the historic building. The project is being developed under the 'Revitalising Historic Buildings Through Partnership Scheme' introduced by theoffice of the Commissioner for Heritage under the Hong Kong Development Bureau.

Shek Kip Mei Estate plays a remarkable role in the public housing history of Hong Kong as it was the first Mark I type public housing to be built here. The project preserves both the tangible heritage (built heritage) and the intangible (the building's usage); it ensures that the operation will be viable and sustainable; and it aims to bring out the value of Mei Ho House through the preservation of its character-defining elements.

Design

The design and installation of building services for the Mei Ho House revitalisation project consider the future ease of operation and maintenance for its operation as a city hostel [Fig. 7a andFig. 7b]. The overall building approach at a design level involves:

- Design for environmental comfort and sustainability. Natural and full-spectrum lighting exposure is maximised, as is passive ventilation in the hostel for a healthy living space and good indoor air quality.
- Communal spaces. Circulation and lobby space shall be designed for interaction and provided with greenery. "Communal pocket spaces", "interaction nodes", and a "common room" will suit the activities of small neighbour groups and the general public.
- Multi-storey hostel special consideration. The design integrates vertical circulation with open corridors and communal spaces as design features for interaction and identity.

Housing Museum as Portrayal of Living History

A housing museum incorporating preserved basic living units as well as the restored façade will enable the public to re-visit the collective memory [Fig. 8aand Fig. 8b]. Thematic exhibitions will be shown portraying the living style of the old days and social transformation. The result will be a community heritage landmark and enhanced public interaction. Suspended flexible display panel systems will be adopted for ease of changing exhibits andto prevent damage to the original walls.

Comfortable Guestrooms

Targeting a three-star hotel status, the guestrooms will provide a comfortable and decent environment [Fig. 9]. The traditional guests of the Youth Hostels Association come from around the world and would enjoy a home-away-from-home feeling at Mei Ho House.

Case Study – Nga Tsin Wai Village Conservation Project

The 'Lost' History

Nga Tsin Wai Village is a walled village in the urban area [Fig. 10]. The conservation approach adopted asserts that under the existing conditions, it is not worthwhile rebuilding the whole village. Rather, a conservation park is proposed.

The design will adopt the innovative concept of "Conservation by Design" to preserve three relics of the village, namely: the village gatehouse, the embedded stone tablet and Tin Hau Temple, and the central axis that links them.These are set as the core design elements for the conservation park in order to manifest the ambience of the 600-year-old village, while residential redevelopments can proceed in parallel.

Case Studies – Summary

The projects demonstrate the value of:

- Connectivity. An interconnected street grid network disperses traffic and eases walking. A high-quality pedestrian network and public realm makes walking pleasurable.
- Mixed Use and Diversity. There should be a mix of shops, offices, apartments, and homes on the site. People of all ages, income levels, and cultures can enjoy themselves without difficulties.
- Quality Architecture and Urban Design. Emphasis is put on beauty, aesthetics, human comfort, and creating a sense of place. Human-scale architecture and beautiful surroundings nourish the human spirit.
- Traditional Neighbourhood Structure. Public open space is designed as civic art. Traditional social networks for the elderly are taken into consideration at the designing stage.
- Sustainability. The environmental impact of the development and its operation is kept at a minimal level through eco-friendly technologies and respect for natural systems.

图8a Fig.8a 公屋博物馆方案室内设计 Interior views of the proposed Housing Museum

图8b Fig.8b 公屋博物馆方案室内设计 Interior views of the proposed Housing Museum

图10 Fig.10 衙前围村及开发方案 View of Nga Tsin Wai Village and proposed development

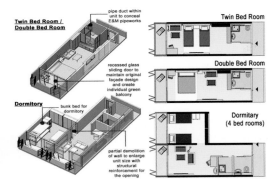

图9 Fig.9 双人间及多人宿舍布局 Layout of twin/double and dormitory rooms

主题Ⅲ：
密度研究1/2
Subject Ⅲ:
the study of density

若阿金·萨巴特·贝尔
加泰罗尼亚理工大学城市与区域规划学院教授

Joaquim SABATÉ BEL
Professor, Urban and Regional Planning, Polytechnic University of Catalonia
Barcelona, Spain

中层高密度
密度的影响程度究竟有多大?
Mid-rise, high-density. To what extent does density matter?

我们应邀根据一个密度为1 000人/公顷（或10万人/平方千米），加上工作场所和设施，探讨城市替代方案来应对巨大的挑战。但是这些要求是否迫使我们修建高层建筑，设计几乎是自主型的巨型建筑？我将对中高层大楼建设方案是否能够满足所需的数量要求，以及这些方案是否更能够为快速老龄化的社会提供一个更友好的环境进行探讨。同时，我还想强调一种理念：当代城市设计必须基于一定程度的合理性和结构秩序。需要更新但永远不能摒弃重复或节奏等传统的城市工具。最后，在"积极养老"和"居家养老"要求的背景下，我将讨论除密度之外还需要考虑的其他因素，以确保城市性甚或城市公平性等的质量。

We are invited to explore urban alternatives to an ambitious challenge, based on a density of 1,000 people per hectare (or 100,000 people per square kilometre), plus workplaces and facilities. But do these requirements force us into high-rise buildings, to the design of more or less autonomous mega structures? I would like to explore whether mid-rise proposals may meet the demanded quantitative requirements and if they are even more capable of offering a friendlier environment for a rapidly ageing society. I would also like to defend the notion that the design of contemporary cities has to be based on a considerable degree of rationality and structural order. Traditional urban tools, like repetition or rhythm, need to be updated but never avoided. Finally, in the context of the "active ageing" and "ageing in place" requirements I would like to argue the need for consideration of other factors besides density to ensure qualities like urbanity, or even urban equity.

在七八十年代的时候，"低层、高密度"已经成为一个流行的话题。大量的论文和研究项目探讨了这一论题，设计和分析了住宅区或城市扩张，试图将这一结合作为应对城市挑战最有效的方法之一。但是一旦有人对其中的任何一种方案进行测量，你就会发现，这种模式的总密度很少超过每公顷六十间住宅，或最多达每公顷七八十间住宅。对此，学者和规划师一点不觉惊讶。对一些城市传统来说，每公顷四十间住宅已经是一个很大的密度了。

亚洲立体城市邀请我们根据一个密度为10万人在每平方千米（或1 000人/公顷）的区域内生活和工作，加上工作场所和设施，探讨替代方案来应对巨大的挑战。这些要求必然需要我们采用高层建筑吗？它们是否引导我们设计自主型的巨型建筑？

恰巧，在收到新加坡国立大学（NUS）邀请参加本次讨论会的邀请后第二天，我的一个好朋友给我送了两份PPT演讲稿，让我指出其中我所偏好的一份。一组图像是关于矗立着精致摩天大楼的快速发展、高密集的城市。另一组图像则描绘了欧洲某个狭小、宁静、舒适的传统村庄。他问了一个有趣的问题：你更喜欢住在哪里？我不得不承认，我坦白地说："……都不喜欢，我更喜欢两者综合的城市——既有高密度和强度，又有城市性、多样性，以及良好的城市景观。"通常，城市规划师重视我们所居住或参观的城市特征。我想说的是，如果采用中层建筑，城市的这些特征可以得到保障。我将尽力证明，并非单一的巨型建筑，或"公园塔"（摩天大楼）组合才能体现这种特征。同样也可以采用基于适当的更新以及众所周知的都市化格局。

中层高密度城市来实现友好环境

一年前，在第一届亚洲立体城市讨论会上，我们意识到，相比竞赛要求，被认为是世界上密度最大的许多城市实际上并不是那么密集。纽约中心区域的平均密度仅每公顷154人，最大密度在曼哈顿中心区，达到每公顷530人。上海中心区域每公顷将近247人，城市中心密度最大，约为每公顷962人，其他被认为是很密集的城市，如墨西哥城、圣保罗或孟买等，实际低于竞赛中所设定的总密度：每平方千米10万人或每公顷1000人。

如果测量限于城市的居住区内，密度当然会高一些。我们考虑一下最著名的一个例子：新加坡的Pinnacle@Duxton[图1]。Pinnacle的容量惊人，每公顷的公寓达600栋。在新加坡的公共住房历史中，它是一个标志性的项目，具有许多独特的特征。它是新加坡第一个楼层

图1 Fig.1　新加坡达士岭租屋密度达到每公顷600户 The Pinnacle@Duxton, a public housing development in Singapore, achieves a density of 600 apartments per hectare

达到50层的公共住房项目，包含7栋摩天大楼。同时，它也是世界上第一栋设有两架天桥的项目，该天桥连接了大厦的第26层和第50层，也确实是一个例外。我将讨论一些更常见的城市景观。

我想讨论测量的另一方面。为此，我将采用一个我所熟知的城市作为案例研究，你们当中许多人可能也知道这座城市，就是巴塞罗那。同其他城市一样，在巴塞罗那，你会发现一些非常传统的城市组织；一座有城墙的历史古城。巴塞罗那的高度和密度一直增长直至19世纪中叶。在过去的40年里，古老街区的面貌和特征发生了巨大变化。居民密度达到每公顷333人，其中将近1/3的都是退休人员，18岁以下的只占1/10。建筑面积比为每平方米3.36平方米。三大主要方向是旅游、第三产业以及象征性和机构中心。

我们也会发现经典城市扩张（扩建区）——这是由Cerdà I Synyer于1859年设计的欧洲最大的城市扩张[图2和图3]。如今，它仍然是巴塞罗那的中心地带。你可以欣赏到它的城市结构、大气的内部庭院以及类似的建筑外观，仅六至八层高。同许多其他欧洲城市一样，巴塞罗那仍存有一些落后城镇，那里有相当高的塔楼和街区，而商业设施和服务所占的比例很低。通常，人们认为这些城区是最密集的区域，可能你们许多人也这样认为，但它们真的那么密集吗？让我们来检验一些例子。Montbau[图4a和4b]是1960年在巴塞罗那北边的一个缓坡上开发的住宅区。长街区和高塔的结合遵循了现代建筑运动的原理，形成奇妙的开阔空间组合，但它的总密度仅每公顷120栋房屋，540位居民。10年之后，在向东不到1千米的位置，建筑师拟议了一项高塔与复杂的混凝土隧道框架的组合[图5]。地形条件以及城市风格方面欠考虑使得结合显得没有意思。即使楼层达到15层，密度也低于前者。

设计Montbau的同一批建筑师为Sundoeste del Besos住宅区[图6]提出了一项有趣的结合。这里地形相当平缓，该方案的特点让我想起了荷兰建筑大师Van den Broek en Bakema设计的一些项目，或柏林的Interbau住宅区。

我们来看一下六七十年代介于Cerdà的城市扩建范围与外围之间的其他住宅区。在某些情况下，甚至采用了开放街区的替代方案。在Viviendas del Congreso[图7a和图7b]的案例中，布局仿效了Clarence Stein与Henry Wright的某些纽约提案（例如：希尔赛德疗养院公寓和菲利普斯花园公寓）。街区与建筑物的接合形成了丰富且复杂的城市图景。半个世纪之后，该住宅还保持着高度的城市风格。家庭成员的高比例使得密度增大至每公顷超过1 000位居民，楼层的高度平均低于六层。其他方案，如用车辆通道将Cerdà街区分割成四部分的方案，每公顷约300栋住宅[图8]。

位于奥运村背后的一个更加精炼的方案，在同一时期修建，结合了Cerdà三条街区[图9]，向我们展示了如何将宽大的中心庭院与地区设施相结合。只有12米高的精致住宅密度高达每公顷175栋住宅。

但让我们回到历史上巴塞罗那（扩建区）的中心城市扩展区，观察三位现代主义大师Antoni Gaudi、Josep Puig i Cadafalch和Lluis Domenech i Montaner 是如何将他们的作品与当代其他建筑结合在一起的。[1]这个令全球钦佩的"Manzana de la Discordia"（"不和谐街区"）[图10a和图10b]实现了每公顷250栋住宅（超过亚洲垂直城市的要求）。我们甚至能够找到每公顷超过500栋住宅的高密集街区。

一些结论似乎已经清晰明确。巴塞罗那扩建区这个传统的城市结构，设计于19世纪中叶，密度高于周边有巨型街区的住宅区[图11a和图11b]。它的密度是上海和纽约住宅密度的两倍。这一城区的非居民活动，如商店、服务业和办公室所占的比例也很大。非居民建筑面积和工作场所的数量与居民住宅面积和居民数量相当。巴塞罗那城市扩张是西班牙最大的集中活动。

最值得注意的是这些数据全部来自于这样的城市结构：街道平缓宽阔、绿树成荫[图12]，及近年来翻新为公共花园的大量家用空地和内部街区庭院。这种城市形态适合于老龄化加快的社会吗？

成为全球老年友善城市

我们应对竞赛要求进行调查：应对和预测快速老龄化社会所挑战的平衡环境城市生活；老年人得到便利、社会关怀和支持的新途径；适合于老一代和年轻一代的建筑环境；以及通过整体分析，鼓励老年人保持独立。一个老龄化的城市还需要具备其他什么具体特征呢？

关于这个问题已经进行了大量有趣的研究，从中我们可以了解到：1. 同任何其他居民一样，老年人也有不同的偏好，但通常他们希望尽可能靠近自己成长的地方或处于一个熟悉的环境，因为他们一直住在那里；2. 建议靠近公共交通车站、副食店和超市，以便能够利用其它城市设施，如餐馆、理发店、保健、各种公园和花园以及特殊的教育和运动配套设施（不超过两个街区或200米）。整个城市的开阔区域应统一分布，间隔不超过500米；3. 预防医疗中心和公共游泳池和体育馆等体育运动区域也应靠近他们的住宅。老年人不希望特殊治疗，因此可以方便地将他们的护理与一般的邻居卫生中心结合在一起；4. 强烈建议在他们的建筑物内布置常见的社交聚会场所（有集体洗衣机、烘干机和叠衣服的存放地方）。老年人喜欢保持独立，但也渴望与其他居民保持密切的联系；5. 各年龄层人们都需要归属于一个特定的地方。他们希望能够清晰地识别自己所居住的与其他区域的不同。通常建议按照大约500人进行分组。6. 为了确保良好的城市质量，同时保持各个年龄的人们之间的良好平衡，需要明确高居住密度和高活动强度；7. 老年人应具有积极性，不仅是针对运动，而且还针对老年人拥有的一套权利，以便他们能够继续参与社会、经济、文化、精神和公民活动。第三代是基本的经济和社会资源。城市为人们提供工作到70多岁的机会，这一点至关重要。因此，工作场所应靠近住宅区和服务区；8. 一些城市正在成为"全球老年友善城市"。这些城市为体育活动提供了公园和开阔的空间。它们促进了社会项目、便利的交通和廉价的住房。同时也提出了靠近居民区和需要部分医疗救助的第三代人口，以及靠近年轻家庭的医疗中心的错位，以便老年人和年轻人都能够保持稳定的关系。

我们一步一步得出结论，即建设一座更加老年友善的城市能够实现双赢。良好的环境对老年人有好处的环境对所有年龄的人们都有好处。实际上，通常来说，我们需要的就是良好城市。那么一个真正良好的城市需要具备什么条件呢？

相当程度的合理性与结构秩序

看起来至少有两个必不可少的条件：精心组织的基础设施和丰富多样的城市风格，也就是说，物质支持和强有力的精神意向。我的假设是为了确保第一个要求，我们需要某种结构秩序。对于巴塞罗那来说，150年前Cerdà的格局就已经规定了这一秩序。

通常来说，城市的设计（或城市扩张）基于相当程度的合理性（不同于绝对规律性）。确保这一合理性的其中一个最常用的工具就是规则网格，但这并不是唯一的工具。

Cornelis van Eesteren设计的阿姆斯特丹城镇扩建项目（AUP）[图13]是20世纪最著名的城市项目之一，它是基于复杂的序列主义随着变化采用对位法和重复方法发展起来的。序列主义这一方法曾经用在绘画和雕塑中，但在音乐作曲家Anton Webern的作品中也有出现。我们知道其他许多特定城市结构形式的例子，这实现了形式和功能的接合。在任何情况下，我们都需要某种结构秩序，即使你仔细看一下去年比赛获奖人员的精美绘画，你也会发现一些基本规律——规则使方案合理。

传统或现代的城市手法，如对称、重复、规律变化、对位或节奏需要持续更新，但为了实现良好的城市，我们不能回避这些手法。请允许我就确保合理性的最常见的选择方案之一——规则网络进行一个说明。

我们发现世界上许多的城市都是基于正交网格。19世纪末，Cerdà方案的反对者们批评该方案："……相当乏味且太过规则"[图14]。于是他们设法使街道弯曲，并引入对角线。旧金山的瑞莱恩斯大厦、芝加哥、碧湖市、马尼拉以及其他许多方案也是一样，都遵循了学院派城镇规划师的原则。这些批评家正确吗？规格网格有必要被非议为绝对统一和及其乏味吗？

请允许我对相反的论点进行辩护："没有比相当规则的

网格更灵活的了。"1811年曼哈顿委员会计划[图15]提供了12条宽度均为300米的南北向大道，155条宽度均为20米的街道，位于哈德顿河和大西洋之间。这些包含了1860个街区，所有长度均为120或240米，宽60米。每一个街区都划分为统一的体块，每一个地块长30米，宽7.5米。结果有超过120000个地块大小几乎一样。委员们提出了一项街道、街区和地块的格局，几乎没有空地、预览设施、没有考虑地形或水道。没有河流或海洋干道、没有公园，也没有特殊的城市中心，仅仅就是一个规则网格。

在报告中，委员们解释称，引起他们注意的首要对象之一就是他们是否应限制在直线街道中，或者是否应该采用一些假定的改进方法，如圆圈或对角线。最后他们得出结论，一座城市必须主要由人居建筑组成，直边和直角房屋是最方便居住的房屋。实践理性和效率在世界上最大的首都之一的设计中得到了体现。看来他们是正确的。1856年的一张图片[图16]表明首都（真实的曼哈顿）得到了快速辉煌的发展，这里工厂烟囱和蒸气船代替了教堂的塔尖和帆船，城市开始了令人眼花缭乱和持续的变化。

两个多世纪的转变[图17]证明了这样一个简单的决定——规则网格所假设的灵活性和不同寻常的多样化。伟大的建筑绘画家和建筑师Hugh Ferriss在他的作品《明天的大城市》[2]中展示了这一点。但是今天的现实仍然超出了他丰富的想象。

在巴塞罗那的东北部也出现了类似的事件，其中22@区的发展看起来像一朵蘑菇，城市和建筑具有很大的灵活性[图18]，[3] 200公顷的地区正在经历快速的转变，其中项目超过了100个。大多数的项目超过半公顷，都已经得到了批准。这些项目创造了40 000个新的就业机会，甚至是在经济危机加剧时。这意味着总住房面积超过两百万平方米，新增了20 000栋住宅。

根据150年前详细阐述的愿景，这一变化仍然受到Cerdà街道格局的支持，如今现代技术的所有制约因素都在支撑这一变化。像在纽约一样，巴塞罗那在1959年也制定了相当简单的规则：街道的对齐、下水道系统的布局、公共交通与公共活动以及建筑物的外观。

如今在22区，大部分的建筑物都是独立支撑。建筑可以变得非常自主，也许这种自主太多了，使得基础设施格局仍然保留，它规定了建筑面积比、活动密度，最重要的是，它提供了一个整洁的城市秩序。

很显然，确保秩序和合理性的方法有许多种，但网格可能是最好的一个方法。当网格伴随着智能建筑规程时，它们就会变得相当丰富灵活。结合越紧密，结果就会越灵活。网格合理性的强度也许最永恒，而且最忠实于所有城市规律的公共规范，支持系统的灵活性使得Passos的魔术山（曼哈顿）[4]或22区成为了可能。这一结合未能使100年前Antoni Gaudi、Josep Puig i Cadafalch和Lluis Domenech i Montaner（三位伟大的现代主义大师）没有遵循规律而是尊重街道的格局成为可能，创作出的许多不同建筑的巴塞罗那"不和谐街区"的著名临街地界。他们创作了其中最有趣的城市和建筑故事。同样的灵活性包括了巴塞罗那更大的网格元素，如Clinic医院、工业学校和圣保罗医院等。支撑的刚度证明了建筑的自由和城市的美丽——"多元合一"的原则。

密度、合理性，而且还是对城市文脉负责的优质建筑

我认为建设一座美好的城市要有两个必不可少的要素：不但要有精心组织的基础设施，而且要有丰富的城市化程度。我将就那些能够确保城市质量，如城市风格，甚至城市公平等进行探讨。

很显然没有万能药。我们需要各种密集的活动（住宅、商业、办公和便利设施）和对城市文脉负责的优质建筑，但这还不够。密度已经成为描述一座城市的形态和质量最相关的属性之一。但我们应该对一些定义和局限性加以评审。去年，我认为我们达成了一个共识，即高密度更具有持续性。20世纪，社会和经济趋势促进了低密度，也产生了资源过度消耗带来的一些新问题（运输、货物、能源和空间）。很显然，单是高密度还不足以解决这些问题。但今天我们普遍承认低密度会导致对私家车的依赖，增加二氧化碳的排放，减少社会接触和城市风格。密度的增大已经成为了实现更加可持续发展城市的一个主要需求之一，是城市规划中的一个明确的重点。我也尽力说明要取得积极的效果，光是高密度还不够，还应随其具备真正有效的公交系统和严格的空地要求。世界上最大和最密集的城市向我们说明了它们对公共交通的依赖程度，或有多少人必须骑车或步行到达自己的工作地点。由于增加了电子办公、商住两用房，以及提供了相当有效的公交系统，因此减少强迫运动量绝对必要——正如我们在提交竞赛的所有方案中所看到的那样。

但现在我想就更广义的密度理解做一个辩护——这个定义不仅涵盖人口和住宅数量的测量，而且还包含人口和住宅的类型，以及他们的活动，或建筑物能够容纳的活动。正如在自然界一样，我们也会谈到城市多样性，并运用其中的一些原则。从这个意义上来说，就是社会阶层、背景和根源的混合、各种建筑类型，以及促成城市熵的不同活动间的平衡。最后，良好的平衡使得城市更宜居，更加可持续。

因此广义的城市性这一概念应包括活动强度的测量或上述特征的种类。我们的朋友Meta Berghauser和Per Haupt开发了一种强大的工具——'空间矩阵'，我们需要继续对该领域进行研究。[5]最后，让我们记住刚刚过世的一个伟大的建筑师和专家的一些想法。如Manuel de Solà-Morales所说，城市性指的是与建筑相关的三样东西：可透性、感官享受和尊敬。[6]可透性就是进出和进入一个建筑物的能力。城市建筑物从物理或精神上讲会提供很多的进入方式。基层是最具可透性的空间（商店、门廊、入口和大门）。也存在其他空间，其中建筑师作为街道的一部分，与城市融为一体。可透性是建筑城市性的一个基本要素。

第二个要求是感官享受。接触建筑物不是多余的；这是城市的基本问题。如果我们想碰一个建筑，如果我们感觉跟它很亲近，且受到它的吸引，而非排斥，我们就能感到快乐。感官享受指的是所有感觉同时都感到愉快。保证城市性的一个重要感觉就是恰当地接触——当我们走在人行道或站在地板上时我们的接触。穿过城市也是理解、了解城市的一种方法。在理查德·森尼特解释人体如何与城市相关时，谈及了这种感官享受。[7]

与所有非常有趣的概念一样，尊敬也是一种模糊的概念。森尼特继续说道"……在不平等社会中，尊敬是维护人的尊严的唯一方法"。尊敬不是敬畏。尊敬不需要遵守普通法律或规则。正好相反，尊敬是强者对弱者、公

图2 Fig.2 巴塞罗那扩建区 Barcelona's Eixample town extension

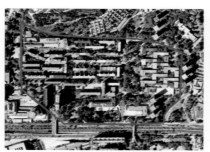

图4a Fig.4a 巴塞罗那Montbau住宅区(1960)密度为每公顷120户540人 Barcelona's Montbau housing estate (1960) reaches a gross density of 120 dwellings and 540 inhabitants per hectare

图5 Fig.5 坎耶亚住宅区（1973）每公顷112户505人 The Canyelles housing estate (1973) offers 112 dwellings and 505 inhabitants per hectare

图7a Fig.7a 国会两院住宅区（1954）每公顷240户1030人 The Viviendas del Congreso housing estate (1954) contains 240 dwellings and 1,030 inhabitants per hectare

图3 Fig.3 扩建区提供宽敞的街区内部庭院 Eixample offers generous inner-block courtyards

图4b Fig.4b 巴塞罗那Montbau住宅区(1960)密度为每公顷120户540人 Barcelona's Montbau housing estate (1960) reaches a gross density of 120 dwellings and 540 inhabitants per hectare

图6 Fig.6 德尔贝索斯西南住宅区（1960）每公顷140户631人 The Sudoeste del Besos housing estate (1960) achieves 140 dwellings and 631 inhabitants per hectare

图7b Fig.7b 国会两院住宅区（1954）每公顷240户1030人 The Viviendas del Congreso housing estate (1954) contains 240 dwellings and 1,030 inhabitants per hectare

共机构对市民的关注。建筑师和建筑在与城市的关系中也应采用同样的态度。因此，引入尊敬这一概念是非常重要的。不应将尊敬理解为是对建筑规程或普通法律的遵守。最后，简而言之，建筑就是构建城市的东西：一个好的或坏的城市、一个美丽的或无序的城市、一个大的或小的城市，都有自己的问题。孤立的建筑，即没有城市化考虑的孤僻的建筑物，不仅仅不利于城市性，也不利于城市。我们知道独立建筑虽然不是解决方案，但却是解决方案的一部分。建筑需要做只有好的建筑才能做的事：以立体方式说明任何位置的复杂性，使自身充实，传递城市内涵和间或向我们展示其历史根源。在建筑遗弃城市时，我们的城市正受苦，这是因为没有城市建筑的城市是死的。我们必须意识到城市不是问题，但却是一个可能的解决方案，总而言之，城市是人类最伟大的发明——我们最宝贵的财富。

'Low-rise, high-density' became quite a fashionable topic in the 1970s and '80s. Hundreds of papers and research projects addressed this issue, designing and analysing housing estates or urban extensions, trying to defend this combination as one of the most efficient answers to urban challenges. But as soon as someone measures any of these proposals you will discover that their gross density rarely exceeds 60, or at most, 80 dwellings per hectare. This fact should not surprise scholars and planners. In several urban traditions, 40 dwellings per hectare is already a significant density.

Vertical Cities Asia invites us to explore alternatives to a very ambitious challenge, based on a project where 100,000 people will live and work on a one-square-kilometre area, or 1,000 people per hectare in terms of density, plus workplaces and facilities. Do these requirements necessarily demand the adoption of high-rise buildings? Should they lead us inevitably to the design of autonomous megastructures?

Maybe just by chance, the day after receiving the kind invitation from the National University of Singapore (NUS) to participate in this symposium, a good friend of mine sent me two Powerpoint presentations and invited me to indicate my preference. One set of images was related to fast-growing, intensively active cities with sophisticated skyscrapers. The second set of images described the scenery of a small, quiet, pleasant traditional village, somewhere in Europe. He brought up a tricky question: Where would you prefer to live?

I have to confess that my answer was "... in none of them, but within an urban context that has a good mixture of both – of high density and intensity, but also of urbanity, variety, and a qualified urban landscape." Normally, city planners appreciate these characteristics of the cities where we live, or that we visit. What I would like to argue is that these qualities can be assured in cities with mid-rise buildings. I will try to demonstrate that they do not necessarily require singular megastructures, or a combination of 'towers in the park' (skyscrapers). They can be based on properly updated, well-known urbanism patterns.

A bid for a mid-rise, high-density city to achieve friendly environments

A year ago, in the first Vertical Asia Cities symposium, we realised that many cities, which are supposed to be among the most dense in the world, are actually not so dense compared with the requirements of the competition. New York reaches an average of only 154 people per hectare in its central area, with a peak of 530 people per hectare in the heart of Manhattan. Central Shanghai rises to nearly 247 people per hectare, with a peak of around 962 people per hectare in its urban core. And other supposedly very dense cities, such as Mexico City, São Paulo, or Mumbai, remain below the gross density established in our contest: 100,000 people per square kilometre, or 1,000 people per hectare.

If we constrain our measurements to residential city fragments, densities can of course be quite a lot higher. Let us consider one of the best-known examples: The Pinnacle@Duxton in Singapore [Fig. 1]. With its impressive volume, The Pinnacle achieves a density of 600 apartments per hectare. It is an iconic project in Singapore's public housing history, with many unique features. It is the first 50-storey public housing project in Singapore, and incorporates seven towering blocks. It is also the first in the world with two unique sky bridges linking the blocks at the twenty-sixth and fiftieth storeys. But it is also quite a singular exception. And I would like to discuss more common and repetitive urban landscapes. Let me present another discussion about measures. For this purpose, I will use as a case study a city that I know well, and that many of you probably also know: Barcelona. As in many other cities, in Barcelona you may find very traditional urban tissues; in this case, an historic walled city. It grew in height and density until the mid-nineteenth century. It is an historic quarter that has drastically changed its physiognomy and characteristics in the last 40 years. Residential density achieves 333 people per hectare, with nearly a third of this number retired, and only a tenth below 18 years. The floor-area ratio is 3.36 square metres per square metre. The three main directional themes are tourism, tertiary activity, and symbolic and institutional centrality.

We will also discover a classic town extension (the Eixample) – the largest in Europe – designed by Ildefons Cerdà I Sunyer in 1859 [Fig. 2 and Fig. 3]. Today it remains the real heart of Barcelona. You may appreciate the quality of its urban fabric, the generous inner-block courtyards, and the relatively homogenous building facades, only six-to-eight floors high.

Like many other European cities, Barcelona also has some post-war suburbs, with towers and blocks of considerable height, although a low proportion of commercial facilities and services. Usually people consider these urban areas the densest ones, and probably many of you would think in the same way; but are they really so dense? Let us examine some examples.

Montbau [Fig. 4a and Fig. 4b] is a housing estate developed in 1960 on a gentle slope in the north of Barcelona. The combination of long blocks and high towers follows the principles of the modern movement, and it provides quite an interesting combination of open spaces. But it reaches a gross density of just 120 houses and 540 inhabitants per hectare.

A decade afterwards and less than one kilometre eastwards, architects proposed a combination of higher towers with a sophisticated concrete tunnel formwork [Fig. 5]. Topographical conditions and a lack of concern for urbanity provoked a less interesting composition. Even with 15-floor towers, density is lower than the previous case.

The same architects who designed Montbau showed an interesting combination of blocks in the Sudoeste del Besos housing estate [Fig. 6]. The terrain is quite plain, and the quality of the proposal reminds me of some projects by the Dutch masters Architectengemeenschap Van den Broek en Bakema, or the Berlin Interbau.

Let us look at other housing estates from the 1960s and '70s that are not so peripheral, but located within Cerdà's town extension. In some cases, they even adopt open-block alternatives. In the case of Viviendas del Congreso [Fig. 7a and Fig. 7b], the layout echoes some of Clarence Stein and Henry Wright's New York proposals (for example, Hillside Homes Apartments and Phipps Garden Apartments). The articulation of blocks and towers creates very rich and complex urban scenarios. This estate maintains, half a century afterwards, a remarkable degree of urbanity. The singular ratio of family members elevates the density to over 1,000 inhabitants per hectare, with an average height of less than six floors. Other proposals, like the one cutting four Cerdà blocks with vehicular passages, reaches nearly 300 dwellings per hectare [Fig.8].

图8 Fig.8　扩展区内4个街区密度为285户/公顷 Four blocks at the Eixample offer 285 dwellings per hectare

图9 Fig.9　奥运村附近的3个街区（1990）密度为175户/公顷 Three blocks near the Olympic Village (1990) offer 175 dwellings per hectare

图10a Fig.10a

图10b Fig.10b　"不和谐街区"（约1900）密度为220户/公顷
The 'Disharmony Block' (around 1900) offers 220 dwellings per hectare

图11a Fig.11a

图11b Fig.11b　扩展区内高密度街区可达500户/公顷
Denser blocks at the Eixample offer over 500 dwellings per hectare

A more refined alternative, a combination of three Cerdà blocks [Fig.9], just behind the Olympic village and built in the same period, shows us how to combine generous central courtyards with district facilities. Elegant dwelling typologies with just 12 metres depth achieve quite a high density of 175 dwellings per hectare.

Let us travel back in history to the central town extension of Barcelona (Eixample), however, and observe how Antoni Gaudi, Josep Puig i Cadafalch, and Lluís Domènech i Montaner – the three great masters of modernism – combine their work with other contemporary builders. The globally admired "Manzana de la Discordia" ("Disharmony Block") [Fig. 10a and Fig. 10b] achieves 250 dwellings per hectare (which is over Vertical Cities Asia's requirements). We could even find quite intensive blocks that exceed 500 dwellings per hectare.

Some conclusions seem to be clear and categorical. A traditional urban fabric, Barcelona's Eixample, designed in the middle of the 19th century with a remarkable vision, reaches much higher densities than peripheral housing estates with huge blocks [Fig. 11a and Fig. 11b]. It doubles the residential density of Shanghai and New York. This urban area has also a significant proportion of non-residential activities, like shops, services, and offices. The non-residential floor space and the number of workplaces are similar to the residential floor space and the number of inhabitants. The Barcelona town extension is the largest concentration of activities in Spain.

The most remarkable fact is that all these figures have been achieved in an urban tissue with gentle and broad tree-lined streets and avenues [Fig. 12], domestic squares, and numerous interior block courtyards recently refurbished as public gardens. Is this kind of urban morphology well prepared to host a progressively ageing society?

Towards global age-friendly cities

We should look into the requirements of the competition: a balanced environment for urban life addressing and anticipating the challenges of a rapidly ageing society; a new approach to accessibility, social care, and support for the elderly; appropriate built environments for both older and younger generations; and encouragement for senior citizens to stay independent through the provision of a holistic approach. What other specific characteristics would be required in an ageing city?

There are numerous and very interesting studies about this question, and so we can learn that:

a) As any other inhabitant, elderly people have different preferences, but usually they want to remain as near as possible to the areas where they have grown up, in a neighbourhood that they already know well, because they have always lived there.

b) Vicinity to public transit stops, grocery stores, and supermarkets, is quite recommended, so as to other urban facilities, like restaurants, barber shops, health care, varied kinds of parks and gardens, and special education and sport complexes (no more than two blocks or two hundred metres). Open areas should be uniformly distributed throughout the city at intervals of no more than 500 metres.

c) Preventive medical centres and areas for physical exercise, like public swimming pools and gymnasiums, should be also near to their homes. Elderly people don't want to have a special treatment, so it is convenient to integrate their care into general neighbourhood health centres.

d) Common meeting places to socialise within their buildings are absolutely recommended (with collective washing machines, dryers, and a protected place for folding clothes). Senior citizens like to maintain independence, but they also desire to live in close relations with the rest of the residents.

e) People of all ages need to belong to a specific place. They want to be able to clearly identify the part of the city where they live, as something different from all the other areas. Grouping at a dimension of around 500 people is commonly recommended.

f) A high residential density and a high intensity of activities are clear requirements to ensure a good urban quality, and also provide a good balance between people of every age.

g) Elderly people should be active, and this refers not only to exercise, but also to a set of rights by the elderly so that they continue participating in social, economic, cultural, spiritual, and civic activities. The third generation is a fundamental economic and social resource. It is essential that cities provide opportunities for people to work well into their 70s. Therefore workplaces should be near residential areas and service centres.

h) Some cities are gearing up to become "Global Age-friendly Cities". These cities offer parks and open spaces for physical activities. They promote social housing projects, accessible transport, and affordable homes. They also propose the dislocation of health centres, close to residential areas and to the population of the third generation that requires partial medical aid, but also close to young families, in order that both can maintain a constant relationship.

So step by step we arrive at the conclusion that building a more age-friendly city is a win-win situation. A good environment for elderly people is good for people of all ages. In fact, what we need to promote are, generally speaking, good cities. What makes a really good city?

A considerable degree of rationality and structural order

It seems that there are at least two essential conditions: a well-organised infrastructure and a rich degree of urbanity; that is to say, a physical support and a powerful mental image. My assumption is that to ensure the first requirement, we need some kind of structural order. In the case of Barcelona, this order was provided 150 years ago by Cerdà's grid.

Normally, the design of cities (or city extensions) has been based on a considerable degree of rationality (not equivalent to absolute regularity). And one of the most recurrent instruments to ensure this rationality has been a regular grid. But it is not the only tool.

Cornelis van Eesteren's Amsterdam town extension (AUP) [Fig. 13], one of the most celebrated urban projects in the twentieth century, was developed based

图12 Fig.12　巴塞罗那一条绿树成荫的大街 A broad, tree-lined street in Barcelona

图13 Fig.13　科尼利斯·范·伊斯特伦的阿姆斯特丹总体扩展计划
Cornelis van Eesteren's Amsterdam town extension plan (1930s)

on a sophisticated serial composition employing counterpoints and repetition with variation. This tool has been used in some paintings and sculptures, but it also appears in the work of the music composer Anton Webern. And we know many other examples with a specific urban syntax, which allows formal and functional articulations. In any case, we need some structural order. Even if you look carefully at the beautiful drawings of the winners of last year's competition, you will discover some basic rules – rules that give rationality to their proposals.

Traditional or modern urban tools, like symmetry, repetition, regular variation, counter position, or rhythm, need to be continuously updated, but we cannot avoid these tools if we want to achieve a good city. Let me argue in favour of one of the most common alternatives for ensuring rationality: regular networks. Thousands of cities all over the world have been founded based on an orthogonal grid. The opponents to Cerdà's plan criticised it as "quite boring and extremely regular" at the end of the nineteenth century [Fig. 14]. They tried to curve the streets and introduce diagonals. So did Daniel Burnham in San Francisco, Chicago, Baguio, Manila, and many other proposals, following the principles of the Beaux Arts town planners. Were those critics right? Is a regular grid necessarily condemned to be absolutely uniform and extremely boring?

Let me defend the opposite argument: "There is nothing more flexible than a quite regular grid." The Commissioners' Plan for Manhattan in 1811 [Fig. 15] provided 12 north-south avenues, all of them 30 metres wide; 155 streets between the Hudson River and the Atlantic Ocean, all of them 20 metres wide. That implied 1,860 blocks, all of them 120 or 240 metres long and 60 metres wide. Every block was divided into uniform plots, each of them 30 metres long and 7.5 metres wide. So the result was over 120,000 plots of exactly the same size. The Commissioners proposed a pattern of streets, blocks, and plots, with nearly no open spaces, nor previewed facilities, nor consideration to topography or watercourses. No river or ocean boulevards, no parks, no special city centres; just a regular grid.

In their report, the Commissioners explained that one of the first matters that caught their attention was whether they should confine themselves to rectilinear streets, or whether they should adopt some of those supposed improvements, like circles or diagonals. They concluded that a city has to be composed principally of buildings for people, and that straight sided, and right angled houses are the most convenient to live in. Practical reasons and efficiency supported the design of one of the greatest capitals in the world.They seemed to be right. An image of 1856 [Fig. 16] shows the fast and splendid development of a capital (Manhattan), where factory chimneys and water-steam vessels substitute the pinnacles of churches and sailboats, a city that begins its vertiginous and continuous change.

The transformation over two centuries [Fig. 17] confirms the significant level of flexibility and the extraordinary degree of variety that such a simple decision – a regular grid – supposed. Hugh Ferriss, a great illustrator and architect, shows it in The Metropolis of Tomorrow. But the reality is still running today beyond his rich imagination. Something equivalent happens in the north-east of Barcelona, where the 22@ District is growing like a mushroom with a great degree of morphological and architectural flexibility [Fig. 18]. Some 200 hectares are undergoing a rapid process of transformation, with more than 100 projects. Most of them are over half a hectare in size, and have been approved. Some 40,000 new jobs have been created, even in the midst of a deep economic crisis. This implies more than two million square metres of gross floor area and nearly 20,000 new dwellings.

And this change is still supported by Cerdà's street pattern – by a vision elaborated 150 years ago, buttressed today with all the conditioning factors of modern technology. As in New York, quite simple rules were also established here in Barcelona in 1859: the alignment of streets, defining the sewage system, public transport and movement, and the profile of the buildings. Today, architecture can become extremely autonomous – perhaps too much – but in the 22@ District, the infrastructural pattern remains and it defines floor area ratios, activities densities, and above all, it offers a clear urban order.

Obviously there are many other tools for ensuring order and rationality, but a grid is perhaps the most powerful one. When grids are accompanied by intelligent building regulations, they become quite rich and flexible. The firmer the support, the more flexible the outcome. The strength of the rationality of a grid is perhaps the most permanent and most faithfully followed public norm of all the urban laws.

This flexibility within a clear support system is what makes John dos Passos' magic mountain (Manhattan) or the 22@ District possible. This combination is what made possible, 100 years ago, the famous frontage of the "Disharmony Block" at Barcelona, where Gaudí, Puig i Cadafalch, and Domènech i Montaner (the three great modernist masters) created quite different buildings with no rule other than a respect for the street. They created one of the city's most interesting urban and architectonic episodes.

It is the same flexibility that un-problematically includes elements of a much larger size in Barcelona's grid, such as the Clinic Hospital, the Industrial School, and the Saint Paul Hospital. The firmness of the support is the proof of architectural freedom and urban beauty – the "unity in variety" principle.

Density, rationality, but also a good-quality architecture that is responsible to its urban context

I sustained that there were two essential ingredients for building a good city: a well-organised infrastructure, but also a rich degree of urbanity. I would like to discuss what could ensure qualities like urbanity, or even urban equity. There is obviously no magic potion. We need (but that is not enough) a dense mixing of activities – residential, commercial, offices, and amenities – and a good-quality architecture that is responsible to its urban context. Density has become one of the most relevant attributes for describing the form and quality of a city. But we should review some definitions and limitations.

Last year I argued that we share a general consensus that high densities are more sustainable. In the twentieth century, social and economic trends promoted low densities and created new problems driven by overconsumption of resources (transport, goods, energy, and space). Evidently, high density alone is not enough to solve these problems. But today we generally admit that low densities contribute to private car dependency, to the increase of carbon dioxide, to less social contact and less urbanity. Increasing density has become one of the main requirements for achieving a more sustainable city; it is a clear priority in urban agendas.

I also tried to show that high density is not enough to achieve positive outcomes; it should always be accompanied by a very efficient public transport system, and by strict open space requirements. The biggest and densest cities in the world show us to what extent they rely on public transport, or how many people have to cycle or walk to arrive to their workplace. So, reducing the amount of forced movements, due to increasing teleworking, mixing commerce and services with housing, and providing a quite efficient public transport system, becomes absolutely essential – as we have seen in all the proposals presented in the competition.

But right now I would like to defend a broader understanding of density – a definition that covers not only the measurement of the amount of people and dwellings, but also the variety of both, and of the activities they do, or those that buildings could host. As in the

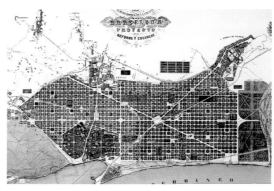

图14 Fig.14　塞达尔1859巴塞罗那扩建规划
Cerdà's plan for the extension of Barcelona, 1859

图15 Fig.15　曼哈顿1811行政首长计划
Commissioners' Plan for Manhattan (1811)

图16 Fig.16　1856年的曼哈顿
View of Manhattan, 1856

图17 Fig.17　今天的曼哈顿
Contemporary view of Manhattan

图18 Fig.18　巴塞罗那22@区
View of Barcelona's 22@ District

natural world, we may talk of urban diversity, and apply some of its principles. In that sense, the mixture of social classes, backgrounds and origins, the variety of building typologies, and the balance between different activities contribute to the urban entropy. In the end, a good balance makes a city more liveable and sustainable.

So this broader concept of density has to include the measurement of intensity of activities or variety of those characteristics. Our friends Meta Berghauser and Per Haupt have developed a powerful tool, "Space Matrix", but we need to continue the research in this field.

Let me finish by remembering some of the thoughts of one of our greatest architects and professors, who recently passed away. As stated by Manuel de Solà-Morales, urbanity means three things in relation to architecture: permeability, sensuality, and respect. We talk about permeability as the ability to access and enter a building. An urban building is one that offers many ways to get in, physically or mentally. The ground floors are the most penetrable spaces (shops, porches, entrances, portals). There are those spaces where the architecture is part of the street, integrated into the city. Permeability is an essential quality for the urbanity of the architecture.

A second requirement is sensuality. The touch of the buildings is not superfluous; it is a basic issue for the city. If we like to touch a piece of architecture, if we feel close to it, attracted by it, not repelled by it, we can reach a feeling of pleasure. Sensuality means a pleasant feeling of all the senses at once. One of the most important senses to guarantee urbanity is precisely touch – the touch we feel when walking on the pavement, or standing on the floor. Walking through the city is a way to absorb it, to understand it. Richard Sennett speaks about this kind of sensuality when he explains how the human body relates to cities.

Like all interesting concepts, respect is also an ambiguous one. Sennett writes that respect is the only way to preserve human dignity in an unequal society. Respect is not reverence. Respect does not meet the general laws or rules. It is just the opposite. It is attention from the strongest towards the weakest, from public institutions to citizens. The same attitude has to be adopted by architects and architecture in their relationships with the city. It is very important, therefore, to introduce the notion of respect. Respect should not be understood as a fulfilment of building regulations, or general laws.

So in the end, architecture is, in short, what builds the city: a good or a bad city, a beautiful or a problematic one, a big or a small city, with all its problems. Isolated architecture, an autistic building without urban consideration, is against urbanity, against the city.

We know that architecture alone is not the solution, but it is part of the solution. Architecture needs to do what only good architecture can do: to express in three dimensions the complexity of any location, to enrich it, to transmit urban meaning and occasionally show us the roots of its history. Our cities suffer when architecture abandons them, because a city without urban architecture is dead. We have to be aware that the city is not a problem, but a possible solution, and above all, the city is the greatest invention of humanity – our most precious treasure.

参考文献 References

1. Antoni Gaudi（1852-1926）、Josep Puig i Cadafalch（1867-1956）和Lluis Domenech i Montaner（1850-1923）是一代最著名的建筑师、艺术家、雕塑家、画家和铁匠，他们将巴塞罗那打造成世界现代主义之都。
Antoni Gaudi (1852-1926), Josep Puig i Cadafalch (1867-1956) and Lluis Domenech i Montaner (1850-1923) are the best well known among an impressive generation of architects, artists, sculptors, painters, blacksmiths, which established Barcelona as the world capital of modernism.

2. 参见Hugh Ferriss《明天的大城市》，纽约：普林斯顿建筑出版社，1986年。1929年版重新印刷。
See Ferriss, Hugh. The Metropolis of Tomorrow. New York: Princeton Architectural Press, 1986. Reprint of 1929 edition.

3. 22@巴塞罗那项目将波布雷诺200公顷的工业用地转变成为创新区域，为知识密集型活动提供了战略集中地。这一首创也是城市改建和新型城市的一个项目，回应了知识社会所提出的挑战。
22@Barcelona project transforms two hundred hectares of industrial land of Poblenou into an innovative district offering modern spaces for the strategic concentration of intensive knowledge-based activities. This initiative is also a project of urban refurbishment and a new model of city providing a response to the challenges posed by the knowledge-based society.

4. 参见John dos Passos,《曼哈顿的转变》，纽约，1925年
See dos Passos, John. Manhattan Transfer. New York, 1925.

5. 见Meta Berghauser和Per Haupt，'空间、密度和城市形式'，博士论文，代尔夫特工业大学，代尔夫特2009年。
See Berghauser, Meta and Haupt, Per, "Space, Density and Urban Form". PhD dissertation, Technische Universiteit Delft. Delft 2009.

6. Manuel de Solà-Morales（1939-2012年）是上世纪最著名的建筑师和城市理论家之一。见http://www.manueldesola.com/curriculum_eng.htm
Manuel de Solà-Morales (1939-2012) has been one of the most important architects and urban theorist in the last century. See http://www.manueldesola.com/curriculum_eng.htm

7. 理查德·森尼特，《西方文明中人体和城市》，WW Norton and Company，伦敦和纽约，1994年。
Sennet, Richard, The body and the city in Western civilization. WW Norton and Company. London and New York, 1994.

DHARAVI CHAMBDA BAZAAR Mumbai, India

The Density Atlas is a planning, design and development resource for comparing urban densities around the world. The Atlas features a unique metrics and scale system for a comprehensive understanding of urban density.

METRICS

FAR   DU/AREA   POP/AREA

Case studies are measured by floor area ratio, dwelling units per area, and people per area, to provide a holistic understanding of density.
Learn More »

SCALE

Case studies are categorized by physical scale in a typical metropolitan region, with a focus on development blocks, neighborhoods, and districts.
Learn More »

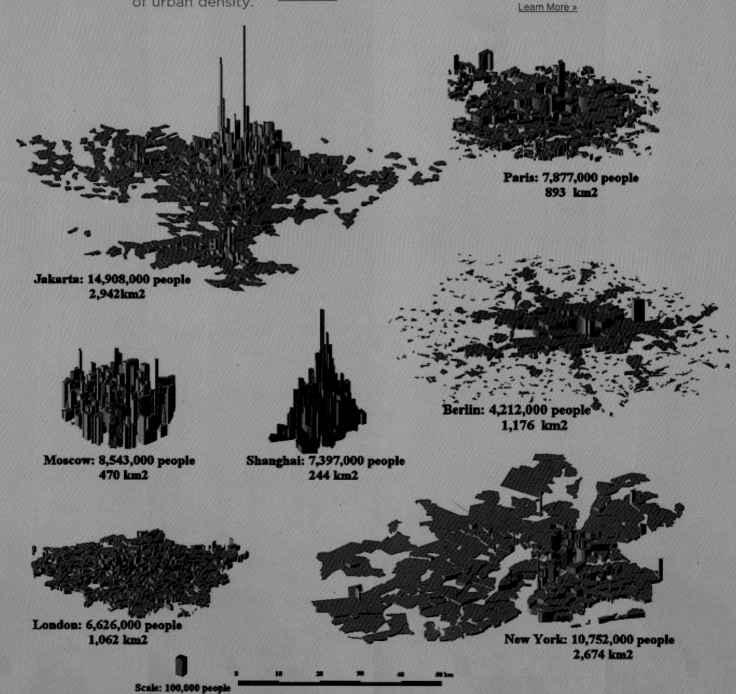

Jakarta: 14,908,000 people
2,942 km2

Paris: 7,877,000 people
893 km2

Moscow: 8,543,000 people
470 km2

Shanghai: 7,397,000 people
244 km2

Berlin: 4,212,000 people
1,176 km2

London: 6,626,000 people
1,062 km2

New York: 10,752,000 people
2,674 km2

Scale: 100,000 people

密度图集
The Density Atlas

主题III：
密度研究2/2
Subject III:
the study of density

李灿辉
麻省理工学院城市研究与规划
系荣誉教授

Tunney LEE
Professor Emeritus,
Department of Urban Studies
and Planning, Massachusetts
Institute of Technology
Cambridge, United States of
America

由于全球城市化已达到了50%，因此密度这一概念对于规划师和建筑师来说就变得日益重要。谈到密度，这就引发了其对宜居性、可持续性、经济因素、犯罪、机遇和多样性的影响的问题。为了帮助测量、理解密度，我们在麻省理工学院进行了一组案例研究，并制作了测量工具，让规划师、开发者、建筑师和学生可以用它对全世界的城市居住密度进行比较。该案例研究和对密度复杂度和认知的调查收录于网站"密度图集"（densityatlas.org）——作为一个不断扩大的资源库，将有助于建筑环境相关决策的制定。

As urbanisation reaches 50% globally, density has become an increasingly important concept for planners and architects. Talking about density raises questions about its impacts on liveability, sustainability, economic factors, crime, opportunity, and diversity. To help measure and understand density, we (at MIT) have created a set of case studies and measuring tools that allows planners, developers, architects, and students to compare urban residential densities across the world. The case studies, as well as investigations into the complexity and perceptions of density, are collected in an online "Density Atlas' (densityatlas.org) - a growing resource that will help inform decisions regarding our built environment.

精确比较需要

计算密度的方法有很多，但不是所有的方法都可用来测量同一个事物。密度是什么？怎样比较密度才准确？密度这一概念激发了人们强烈的观点：积极的和消极的。这太密集了；那不够密集；它很拥挤；它很堵塞；它是蔓延开来的；这里在无序地扩张。我们常常采用"高密度"、"中密度"和"低密度"，但它们是地方自主决定的。每个城市对高密度和低密度都有自己的理解，很难跨市或跨区比较这些概念。

对高密度的推动并不是世界公认的。但是目前（尤其是在学术界）人们倾向于将经济指数放在更高的密度上。爱德华·格莱泽在他的论著《城市的胜利》和文章《摩天大楼如何拯救城市》中，通过经济分析表明在高密度的城市中，所有的指数都会增长。但是他在书中并未给出一个数字表示他所指的"高密度"。大卫·欧文在其所著的《绿色都市》一书中则认为纽约市比郊外更加节能。他指出：在美国，曼哈顿是最节能的城市，主要是因为交通运输。我通读了欧文的这本书，但未能找到任何能量化密度的数据。

交通运输、能量和资源利用都与密度有关，但它们的密度是多少呢？在此也存在对宜居性和生产力的关注，反对密度的人总是会看到犯罪和不健康的环境。根据测量目的以及测量地点，有很多测量密度的方法。在美国，每英亩居住单元（DUs）这一概念十分普遍。建筑容积率(FAR)或（常称之为）地积比率用作规划目的，每平方英里人口密度可以估量人口，也常作规划之用。有时我们把密度和高度联系在一起。格莱泽所说的摩天大楼指的是高度，而不是其他任何东西。这还存在一个覆盖范围问题，你是否覆盖了整个现场？我们将如何以对我们有用的方式来组织所有的这些措施呢？存在着很多规模的测量密度，而且我们必须清楚需要我们测量的是哪类规模。是街区规模、邻里规模、城市规模、地区规模还是城市化地区规模？所有这些规模具有不同的密度；当一种层次转变为邻里规模时，街区规模的居住密度会下降。

当你使用一种规模度来测量另一种规模时，比较起来

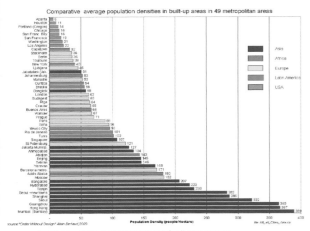

图1 Fig.1
49个大都市建成区平均人口密度比较(Alain Bertaud)
Graph by Alain Bertaud showing comparative average population densities in built-up areas in 49 metropolitan areas

会非常困难。人们利用测量来宣传某个特定的观点，而且定义变得令人困惑。我们使用的方式是试图使密度测量系统化。

研究现状

此项研究执行的部分原因是我在教授综合性设计课程时，我的学生告诉我他们想要创建一个中密度的住宅小区。我就问："那是什么意思，你们想设计多少个单元？"这个问题很难定义。我们发现室外有很多材料，但很难将它们配合起来使用。然而伦敦经济学院关于城市的"城市时代"系列在密度比较上做得很不错。

研究现状表明大部分城市和地区的规模都和建筑师所设计的相反。对城市规模的研究很多，但对社区或街区规模的研究却很少。准确说来社区和街区才是我们建筑师、规划师和城市设计师的工作内容。我们能感受到社区和街区规模的密度。

城市规划师阿兰·柏图针对大都市区做了大量的研究。他比较了众多城市建成区每英亩的人口密度，并在他的网站(alain-bertaud.com)中用图表[图1]展示了他的发现。孟买和香港的人口密度是一个极端，每公顷为300-400人，而亚特兰大则处于另一个极端，每公顷有6人。城市密度大小对比高达60倍。该数据是10年前的数据，比如说对上海就不准确。就人口密度而言，亚洲在一个极端，美国在另一个极端，而南美洲和欧洲则处于中间位置。柏图的图表让我们更好的了解到全世界的平均城市人口密度。

柏图还将大都市区用同样的比例具体化[图2]，并指出了大都市区的高度、宽度和模式，例如将伦敦和柏林同上海、雅加达和纽约比较。伦敦经济学院"城市时代"项目针对不同地方——南美、香港，最近是在印度——的密度布局做了一系列的研究。研究人员还计算了在不同城市中所谓的密集区域的一平方公里中的大部分密度布局，并以图形呈现，他们这项工作做得很不错。

在社区和街区规模方面，航拍摄影师亚历克斯·麦克莱恩同朱莉·波利一起用相机将几乎全美国每英亩的单元拍摄了下来。该研究将数据同社区和街区规模的对接，这是一大开拓。但这种方法也存在一些问题，我会在后面谈及。虽然这种方法在美国运行十分正常，但当你试图用同样的方法测量其他地方的时候，这就变得更困难了。

密度图集

我们在麻省理工学院学院从事密度图集工作很多年。

我们的重点是社区规模，辅以区级和街区规模，试图将该项目系统化，并同密度的测量一致。这是一项需要不断研究的项目，我们希望提供的是一个学习网站，也就是说使该网站充满活力，我们已经发现一些缺陷并正在修改。目前我们大约有100个案例研究，我们想要开放网站获取更多的案例，希望大家能帮助我们进行更多的案例研究。我们的工作已经激发了很多精彩的见解和观察结果，我们打算将这些添加到工作中去。

该项目正在进行中。我们从熟悉的城市波士顿、纽约和香港开始工作。城市的清单在增加，但我们将错过这个世界的一大部分城市，我们还遇到了一些质控问题。此外，还需完成除了测量以外的更多密度方面的工作。

对我们而言，分解密度概念的一个方法是从度量、感知和规划的角度考虑[图3]。度量或测量是我们目前所努力的方面。定义密度主要有两种测量——单位面积人口密度和单位面积建成空间。显然感知在密度中十分重要。每个人对密度都有着不同的观点；价值体系、感觉和舆论也对每个人的密度观点有影响。规划含义对未来可能是最为重要的。那么交通运输、服务、健康、宜居性和能源利用的含义是什么呢？

度量

我们首先来看一下度量[图4]。度量涉及到很多因素。我们的重点主要集中在最广泛的因素。我们认为人口密度和建筑密度是两大关键因素，必须一起测量。只用一种或另一种都会导致错误。首先要确定的是所测规模是城市规模、地区规模、街区规模或社区规模？密度的比较须在同一规模上进行——街区对街区、社区对社区、地区对地区。

我们将一平方公里的区域作为起点，希望涉及城市规模和地区规模。我们的社区和普查区案例研究大部分都在一平方公里的范围内。一平方公里大约是一个城市居民生活所涉及到的最大区域。步行距离界定为400米，大约花10-12分钟，这是我们所能理解的一个规模。社区规模可能为500米乘以500米的平方水平以上。最接近住宅小区规模的面积为一公顷。地区通常覆盖的面积为5 000米乘以5 000米。

一个社区的面积一般约为25公顷和25公顷以上。对"社区"的定义很宽松，有时候用边界来定义，有时候用因素的内部凝聚力（如族群）来定义。在美国"普查区"是一个普遍的概念，很多测量都是根据"普查区"进行，数据也是现成的。人口密度可以在所有的规模中测量。但是为了方便和可行性，建筑密度应在街区层次和社区规模中测量（如果可能的话）[图5]。

因此，测量密度时，我们为什么要弄对规模呢？曾经有个研究声称洛杉矶比纽约的密度更大。如果你把洛杉矶的城市化区域（按照美国政府界定的）同纽约的城市化区域相比较，这完全正确。也就是说，如果你测量这两大区域，纽约每平方英里有5 000人，而洛杉矶有7 000人，但是我们都知道或觉得洛杉矶比纽约的密度小。如果你转到城市水平并根据普查区测量密度的话，你会发现，洛杉矶没有一个普查区的人口密度能达到纽约市2/3普查区的人口密度。如果你着眼于每个普查区，你会发现洛杉矶人口密度最大的一个普查区只有曼哈顿人口密度的一半。

我们的案例研究发现，测量低层/高密度环境、高层/低密度、高层/高密度环境是可行的。接下来我们将用一些例子来说明这一点。首先是孟买的贫民窟，它的容积率大约为2，每公顷大约有3 000人（请注意这些不是准确的数字）。曼哈顿普查区115 001（第5大道第84街）的容积率为5，每公顷大约有444人。很显然，人们有更多的空间。将军澳的高层健明邨和香港的容积率为5.5，每公顷大约有2 550人，人口和建筑密度都很高。

单位面积的居住单元数和人均建造面积也是常用的测量[图6]，二者相互联系。人均建造面积的测量常常用于保障性住房和非正式定居点，有时按照每间房的人数测量。但是人均空间差异极大。你可以想象一下最差的环境，也许是纽约市的廉租房，人均空间可能有2平方米。反之，豪华的居住单元人均空间可能有100平方米。

值得注意的是下东城1900年左右的环境，那时，个别街区的容积率大约为5.3，人口密度为3 700。如今的密度大约为700。该街区的建筑差不多都是一样的。下东城大部分都保持原貌，但是这个区域被贵族化了。不一样的原因自然是由于20世纪90年代初期，每层容纳了2套公寓和16人。而现在由于贵族化以及地区日益繁荣，每层楼只有4人。居住单元一样，但人更少了。现在该街区的容积率为5，人口密度为700，这是个巨大的改变。当我们谈及高密度和低密度的容积率时，我们得到的是完全不同的结果，注意到这一点十分重要。

每个居住单元的空间量极其多变[图7]，所以居住单元也不好测量。595平方米的豪华单元就相当于4个146平方米的单元或是16个37-40平方米的小单元。后者在香港很普遍，在纽约称其为"微型阁楼"。

建筑高度和覆盖范围也是个变量，我们应考虑到容积率[图8和图9]。覆盖范围增加，容积率保持不变。反之亦然，如果覆盖范围保持不变，容积率增加（因为高度增加了）。再次谈到孟买贫民窟的例子，它的覆盖范围是95%，高度为2-3层。将军澳和香港的覆盖范围是40%，高度为39层。曼哈顿普查区115 001的覆盖范围非常高，它的高度为14层，然而其容积率仍然为5。因此我们要考虑所有这些变量。

然而另一个需要考虑的因素是建筑技术，因为它与人息息相关。人们生活的地区生产总值和居民的社会经济地位尤为重要，因为它们会影响对高层建筑的建造、管理和维护能力。

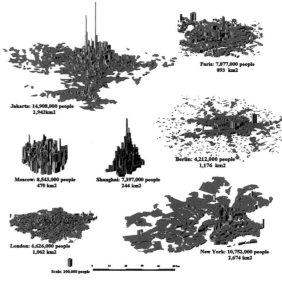

图2 Fig.2 都市结构研究 (Alain Bertaud)
Alain Bertaud's metropolitan structure studies

感知

如前面的章节所讨论的,测量事物并将它们适当地联系起来能使我们了解密度究竟意味着什么。然而感知是非常主观的[图10]。每个人都可以做出判断。有些东西对某人来说很稠密,但对另一个人而言却不够稠密。一个人的观点会受文化因素、价值体系和社会经济环境的影响。

作为专业人士,我们当然有我们自己的观点,但你所处的位置不同,高度也就不同。在我所在的城市坎布里奇市,高度这一概念,也就是高层建筑是令人厌恶的。建筑物高的情况下,人们就会反对"不能有高层建筑,这不是个好事,对你没好处"。另一方面,香港很多人对高度的看法则很积极。"我想要住高点,视野更好,光线和空气也更好"。在上海也是如此。

乔纳森·弗里德曼在他的著作《拥挤和行为》中谈到拥挤对人的影响既不好也不坏,它只是增强了个人对处境的典型反应。人的观点也同社会经济环境有关。20世纪50年代,有些人认为香港密集的公共住房很恶劣,但对于其他人来说,一个带一间卧室的30平方米公寓就能改善他们的居住条件。

规划

对于我们来说,为了更好地规划城市,密度图集也是一个研究密度及其含义的工具。密度如何影响规划呢[图11]?交通运输、服务、健康、宜居性和能源利用的含义是什么呢?这是我们将开始组织的一个领域。建筑形式、城市形态和迁移率会受密度影响——你需要开车吗?步行呢?还是骑自行车呢?公共和私人服务供应也会受到影响。密度越大,所需的服务就越多。什么时候投入电车?什么时候投入公共交通?什么时候运行铁路系统?所有这些东西都和密度有关。

我们希望为人们创造更好的城市——更多产,更高效,并且能源资源能得到更好的利用。我们刚开始探索该话题,仍有很多的工作需要去做。

下一步

[图12]呈现的是密度组成部分以及相关因素的透视图。下一步我们将看一下组成部分之间的特殊关系。密度图集是一个正在进行的项目,其目的在于更好的理解密度这一概念,并用各种方法测量它,帮助人们做出很好的未来决策。现在我们正在制定线路图,以理解各方面的密度以及他们同感知和规划的关系。我们很可能会去除难以理解的"居住单元"分类,让事情简单化。但是每公顷的人口密度和容积率分类可能才是关键问题。

立体城市观测

我们在制作密度图集期间,对立体城市进行了数次观测[图13]。超高密度的立体城市数量很少,首尔和香港就是例子。在垂直城市中,人均空间随着收入的增加和对光线以及空气的要求提高而增加。

例如在深圳,人们对光线和空气的要求提高了,这可能会影响建筑高度,降低密度。有些城市不应建造高楼,因为高楼对建筑技术、管理、公共卫生设施、城市基础设施和服务的水平要求很高。有些国家(比如巴基斯坦、尼日利亚)的经济发展了,但紧迫性的问题是他们如何在没有惨痛剧变的情况下进行转变,改善他们的生活水平,这需要我们以此方法制定合理的发展策略并逐步推进。

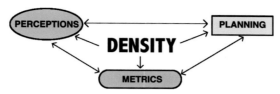

图3 Fig.3 密度组成部分 Components of density

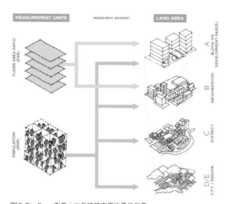

图5 Fig.5 衡量人口及建筑密度的最佳尺度 Scales at which population and building densities can be best measured

Dwelling Units/Area

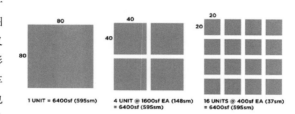

图7 Fig.7 单位面积居住单元数的极端变化 Variations in DUs per area can be extreme

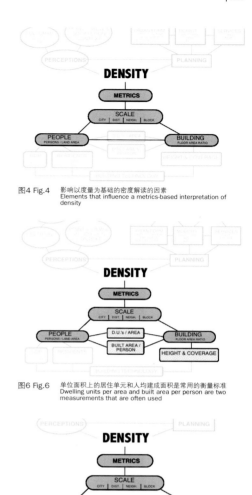

图4 Fig.4 影响以度量为基础的密度解读的因素 Elements that influence a metrics-based interpretation of density

图6 Fig.6 单位面积上的居住单元数和人均建成面积是常用的衡量标准 Dwelling units per area and built area per person are two measurements that are often used

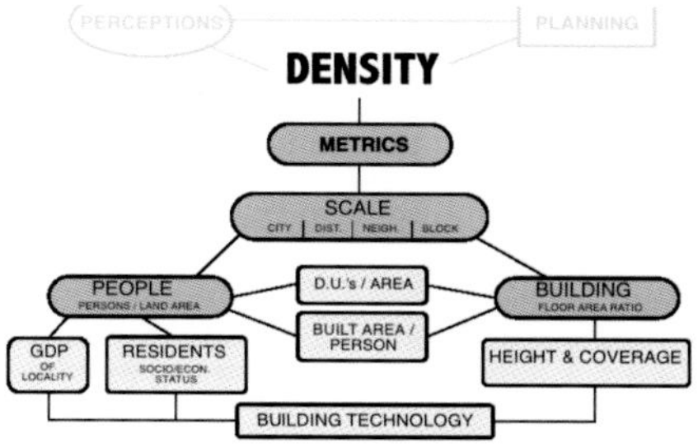

图8 Fig.8 容积率应考虑建筑高度和密度 Building height and coverage should be taken into account with regard to FAR

Need for Accurate Comparisons

There are many ways to calculate density, but they do not all measure the same thing. What is density? How can densities be accurately compared? The very notion of density arouses strong opinions, both positive and negative. This is too dense; that's not dense enough. It's too crowded; it's exciting. It's congested; it's sprawl. We frequently use the terms 'high', 'medium', and 'low' density, but these are locally determined. Each city has its own sense of what it means by high density and low density, but it is very difficult to compare those conceptions across cities or regions.

The push for high density is not universally accepted. However, there is now (particularly in academic circles) a desire to connect economic indices to high density. In his book Triumph of the City and his article 'How Skyscrapers Can Save the City', Edward Glaeser uses economic analysis to show that in high-density cities, all the economic indices are higher. The book Green Metropolis by David Owen argues that New York City is more energy efficient than the suburbs. The most energy-efficient city in the United States is Manhattan, he says, mostly because of transportation. Transportation, energy, resource use – all these are linked to density, but how much density? There's also concern about liveability and productivity, and the opponents of density always look at crime and unhealthy conditions.

There are many ways of measuring density, depending on the purpose and the particular place. The concept of dwelling units (DUs) per acre is very common in the United States. Floor area ratio (FAR) or plot ratio is used for planning and development purposes. People per square mile is used for regional planning. Sometimes we associate density with height.

How do we organise all of those measures in a way that can be useful to planners and architects? There are many scales at which density is measured, and we have to be clear which one we are using. Is it the block scale, the neighbourhood scale, the city scale, the district scale, or the scale of the urbanised area? All these have different densities; the residential density at the block level is higher than the neighbourhood level; in turn, that is higher than the district level. When you use one to measure the other, it makes comparisons very difficult. People use measurements to promote a particular point of view, and definitions become confused. Ours is an attempt to systematise the measurement of density.

Existing Studies

This study came about partly because I was teaching design studios and students would tell me they wanted to create a medium-density development. I would ask, "What does that mean? How many units do you want to put in? How many people will live there?" It was very difficult to define. We found that there was a lot of material out there, but it was difficult to use it together. Existing studies are mostly at the opposite scale to which architects often work: city and district. There have been many studies at the city scale, but few at the neighbourhood or block level. Neighbourhoods and blocks are where we architects, planners, and urban designers do much of our work. We experience density at the neighbourhood and block scales.

Urban planner Alain Bertaud has done a great deal of study on metropolitan areas. He has compared the people per hectare in built-up areas for a large number of cities. He presents his findings graphically [Fig. 1] on his website (alain-bertaud.com). At one end of the scale are Mumbai and Hong Kong in the range of 300 to 400 persons per hectare. At the other end is Atlanta at six people per hectare. City densities vary by as much as 60 times. This data is now ten years old, and is probably no longer accurate for Shanghai, for example. However, as a comparative tool, it works. Asia sits at one end and America sits at the other, with Latin America and Europe in the middle. Bertaud's graph is a good way to get a sense of the worldwide average city population densities. Bertaud has also visualised the metropolitan areas at the same scale [Fig. 2], indicating the heights, breadths, and patterns of, for example, London and Berlin compared to Shanghai, Jakarta, and New York.

The London School of Economics' 'Urban Age' programme has done a series of studies on density layouts in different locations – South America, Hong Kong, and most recently India. Researchers have also taken one-square-kilometre chunks of the so-called densest areas of different cities and assigned figures to them. This work has been very well done.

At the neighbourhood and block scale, aerial photographer Alex MacLean worked with Julie Campoli to photographically visualise units per acre for almost all of the United States. This study was a pioneer in linking the numbers to the neighbourhood and block level. But there are some problems with this method, as I will talk about later. Although it works quite well for the United States, when you try to use that same measure elsewhere, it becomes more difficult.

Density Atlas

We have been working on the Density Atlas (densityatlas.

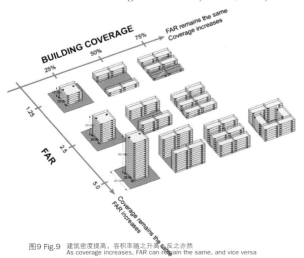

图9 Fig.9　建筑密度提高，容积率随之升高，反之亦然
As coverage increases, FAR can remain the same, and vice versa

org) at MIT for a few years. The project is an attempt to be systematic and consistent with regard to measuring density, and to concentrate on the neighbourhood level, supplemented by the district and block levels. It's a project of continuous learning. We hope to offer a learning website – that is, having made the site live, we've already found flaws and are revising it. We hope that people will help us with more case studies. We want to make it open for additions. We currently have around 100 case studies. Our work has engendered many wonderful insights and observations, and we intend to add these to the work.

The project is a work in progress. We started with cities that are familiar to us: Boston, New York, Hong Kong. The list of cities has grown but we're missing big chunks of the world, and we have some quality-control problems. Furthermore, there is more work to be done related to the other aspects of density besides measurement.

For us, a way to break down the concept of density is to think of it in terms of metrics, perceptions, and planning [Fig. 3]. Metrics, or measurement, is what we've been working hard on. Density is defined by two main measures – people per area of land, and built space per area of land. However, it's clear that perceptions are very important in density. Everybody has a different view of density; value systems, feelings, and opinions come into play. The implications of planning are probably the most important for the future – what are the implications for transportation, services, health, liveability, and energy use?

Metrics

Let us look first at metrics [Fig. 4]. There are many factors involved. We have focussed primarily on the broadest factors; in our view, the density of people and the density of buildings are the two key factors and these must be measured together. To use only one or the other is misleading. The first thing to decide is the scale at which one is measuring; is it the city, the district, the block, or the neighbourhood? Densities have to be compared at the same scale – block to block, neighbourhood to neighbourhood, district to district.

We used the area of one square kilometre as our starting point. We want to relate to the city and district scale. Most of our neighbourhood and census tract area case studies belong within the square kilometre. One square kilometre is about the largest area that an urban dweller can relate to. It can be defined as a 400-metre walking distance, which would take about 10 to 12 minutes. It's a scale that we can comprehend. The neighbourhood scale is probably more at the 500-metre square level. One hectare is closest to the scale of a development. Districts generally cover a 5,000-metre by 5,000-metre area.

A neighbourhood is generally somewhere in the vicinity of 25 hectares. 'Neighbourhood' has many loose definitions; sometimes neighbourhoods are defined by boundaries, and sometimes they're defined by elements of internal cohesion (like ethnic groups). The 'census

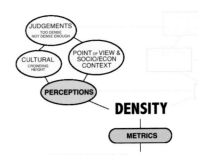

图10 Fig.10　观感是非常主观的密度元素
Perceptions are a very subjective component of density

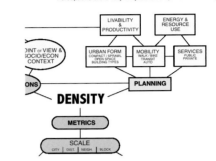

图11 Fig.11　密度效果是重要的规划考虑因素
The effect of density on planning is an important consideration

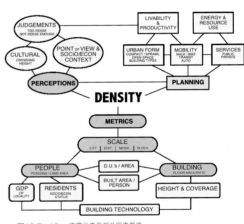

图12 Fig.12　密度元素及相关因素概览
An overall view of components of density and the related factors

tract' is an American concept that is very convenient; a lot of measurements have been made over the entire United States in terms of 'census tracts', and the data is readily available. People density can be measured at all scales. But building density, as a matter of feasibility, is measured at the block level and (when possible) at the neighbourhood level [Fig. 5].

So, when measuring density, why do we need to get the scale right? There is a study that asserts Los Angeles is denser than New York. If you compare Los Angeles' urbanised area (as defined by the US government) to New York's urbanised area, this is absolutely correct. That is, if you measure those two areas, New York accommodates 5,000 people per square mile whereas Los Angeles accommodates 7,000. We all know, or feel, that Los Angeles is not denser than New York. And if you go to the city level and measure densities by census tract, you find that none of the census tracts in Los Angeles come close to two thirds of the census tracts in New York City. If you look at each census tract, the densest (which is the LA census tract), is half that of Manhattan's.

We've discovered from our case studies that it's possible to measure low-rise/high-density conditions, as well as high-rise/low-density and high-rise/high-density conditions. Some examples will provide illustration of this point. Firstly, the Dharavi slum in Mumbai: it has a

FAR of about two, and around 3,000 people per hectare. (Note that these are not exact figures.) In Manhattan, census tract 115001 (at 5th Avenue and 84th Street) has a FAR of five and 444 people per hectare. Clearly, people there have a lot more space. In the high-rise Kin Ming Estate in Tseung Kwan O, Hong Kong, the FAR is 5.5 and there are approximately 2,550 people per hectare – a high density of both people and buildings.

The number of dwelling units (DUs) per area and the built area per person are measurements that are also often used [Fig. 6]. They are interrelated. The measurement of built area per person is most often used in reference to low-income housing and informal settlements – sometimes as persons per room. Space per person can vary tremendously, however. If you take the worst condition you can think of – perhaps tenement housing in New York City – you may have two square metres per person. Conversely, a luxury unit may offer 100 square metres per person.

The condition of the Lower Eastside around 1900 is another example. Back then, a particular block had a FAR of around 5.3 and a population density of 3,700. Today, the FAR is about the same but the population density is much lower. The reason for the disparity is that in the early 1900s, each floor accommodated two apartments and 16 people. Today, with gentrification and increased prosperity, there are four people per floor. Today, a similar block still measures a FAR of five but a population density of only 700. That is a tremendous difference. It is important to note that when we talk about the FAR with a high-density and a low-density, we can get completely different results.

DU's are not always a good measure either, because the amount of space for each DU can be extremely varied [Fig. 7]. A luxury unit of 595 square metres could equal four units at 148 square metres or 16 very small units of 37 to 40 square metres. The latter are common in Hong Kong and called are 'micro-lofts' in New York.

Building height and coverage are also variables we should take into account with regard to FAR [Fig. 8 and Fig. 9]. As coverage increases, FAR can remain the same. Vice versa, as coverage remains the same, FAR can increase (as height increases). Referring again to the example of the Dharavi slum in Mumbai, coverage is 95 per cent and height is two-to-three storeys. In Tseung Kwan O, Hong Kong, coverage is 40 per cent and the height is 39 storeys. And in Manhattan's census tract 115001, the coverage is very high and the height is 14 storeys, while the FAR is still five. We want to take into account all of these variables.

Yet another factor to consider is building technology as it relates to people. The GDP of the locale in which people live and the residents' socio-economic status are very important because they affect the ability to build, to manage, and to maintain high-rise buildings.

Perceptions

As discussed in the preceding section, the ability to measure the various components of density and relate them properly allows us to gain some understanding of density. Perceptions, however, are very subjective [Fig. 10]. Everyone makes judgments. Something can seem too dense to one person but not dense enough to another. One's viewpoint can be affected by cultural considerations, value systems, and socio-economic contexts.

The local context is important. For example, the attitude towards height: In my city – Cambridge –high-rise is anathema. "No high-rise. It's not a good thing. It's bad for you." People would oppose it simply because it is over five stories tall. On the other hand, many people in Hong Kong consider height a very positive thing. "I want to live up there. The view is much better, and the light and air are much better." The same is true in Shanghai.

In his book Crowding and Behavior, Jonathan Freedman talked about how crowding has neither good nor bad effects on people; it rather serves to intensify the individual's typical reaction to the situation. Point of view is also related to the socio-economic context. Hong Kong's dense public housing of the 1950s was looked upon unfavourably by Westerners. But for the residents, a 30-square-metre one-bedroom apartment would have provided an improvement to living conditions and a secure base for family growth.

Planning

For us, the Density Atlas is a tool for learning about density and its implications in order to plan better cities. How does density affect planning [Fig. 11]? What are the implications for transportation, services, health, liveability, and energy use? This is an area we are just beginning to explore. Density affects building type and urban form, as well as mobility – do you need to drive? Is it walkable? Is it bikeable? The provision of public and private services is also affected; the more density, the more services are available. When can you put in a bus system? When can you run a subway system? All these things are connected to density.

We wish to make cities better for people – more productive, more efficient, more enjoyable and with a better use of energy resources. We are just beginning to explore this area, and there is much work to be done.

Next Steps

Figure 12 presents an overall view of the components of density, and the related factors. Our next step is to look at the particular relationships between the components. The Density Atlas is a project in progress. Its aim is to better understand the concept of density and to measure it in ways that can help people to make good decisions for the future. Now we are in the process of laying out a road map for understanding the various aspects of density and their relationship to perception and planning. We are also in the process of changing the layout of the website. We want to make things simpler by (most likely) eliminating the difficult 'dwelling units' category; the people-per-hectare and FAR categories are probably key areas.

Observations of Vertical Cities

We have made several observations with regard to vertical cities during our work on the Density Atlas [Fig. 13]. Vertical cities with hyper densities are few in number. Seoul and Hong Kong are examples. In vertical cities, space per person rises with income, along with the desire for light and air. As incomes rise in Shenzhen, for example, we will see a greater desire for light and air, which may affect building heights and lower the density. There are cities that should not build high-rise housing, because it requires very high levels of building technology, management, public health, city infrastructure, and services. As countries such as Pakistan and Nigeria develop economically, the pressing problem is how they would transform without traumatic upheavals while improving their living standards. That requires a step-by-step process with a strategy based on accurate measures of density.

图13 Fig.13　densityatlas.org

巴黎雷阿勒商场：深层建筑与高层建筑
Paris Les Halles: Deep-rise building versus high-rise building

**主题 IV：
案例分析 1/4
Subject IV:
project discussions**

大卫·曼金
法国Seura Architects
Urbanistes事务所建筑师和城市规划师

David MANGIN
Architect and Urban Planner
Seura Architectes Urbanistes
Paris, France

蒙特利尔、新宿区和洛克菲勒中心都拥有著名的地下城市系统。巴黎雷阿勒商场是"大巴黎"计划的重要交叉站，同时也是一个商业中心、共同娱乐场所、占地面积3公顷的花园、地下交通交汇处等等。本文将对其改建和复杂性进行讨论，为正在积极讨论的"大巴黎快运"环形路线的未来的主要站点提供经验教训。

Montreal, Shinjuku and Rockfeller Center have famous underground urban systems. Paris Les Halles is the main cross station of the 'Grand Paris' plan, but also a commercial centre, a place of public amenities, a garden of 3 hectares, an underground traffic interchange ⋯ This paper discusses its renovation and its complexity as a lesson for the future main stations of the "Grand Paris Express" ring train route, which is being discussed at the moment.

内部城市

在上世纪60年代末，世界各地的各个城市都已尝试了地下城市化，最主要的有蒙特利尔、东京、大阪和规模较小的纽约[图1]。虽然每个城市的情况都有所不同，但他们却有一个共同的因素：都存在地下运输网络，这似乎是发展上述地下空间的条件和动力。正是它产生不可或缺的流动，从而使这些地方（并不会自然地吸引人群）变得有活力。

地下空间因而首先表现为旅行线路。这个线路通过地下站点与地面连接或与另一地下站点连接。在蒙特利尔，地下运输网络的入口都由住宅建筑物的大厅组成。从这里，可以穿过地下连接部分到达商业购物中心。同时，也可以从地下与前一个建筑物地下室相连的毗邻建筑物进入地下运输网络。这一模式将继续下去，直到相距750米远的两个站之间通过连接到城市中心建筑物地下的走廊完全在地下进行连接。

在东京，环状山手线环绕城市中心。市郊列车线路、地铁网络和山手线之间都通过商业购物中心（如新宿区）进行连接，但其很难吸收较高的行人交通流量。对地下运输流的依赖在巴黎雷阿勒商场表现得尤其明显，在这里商业和流动之间的关系完全是颠倒的。这里已经有活动和人群，且商业能产生流动——直到该区域饱和。商业广场是运输的一大分支，且约60%的游客都是通过其地下车站来到这里。因此，地下运输网络的中断能会与地面交通一样导致比较直接的后果。

作为旅程的第一站，地下空间同时也可以作为目的地。巴黎雷阿勒商场的情况就是如此，有一半的游客都会回到他们所来的地方。其中相当数量的游客都是乘坐地铁或快速轨道交通网络，而不用地面交通。同样地，在蒙特利尔也是这种情况，在加拿大漫长的冬季期间，内部城市比地面更受欢迎。但是，在东京，情况就相对有所不同，商业是地下走廊的伴随物，且在一定程度上，也因为这项功能而受到轻视。

基础设施和上层建筑的空间类型有哪些？

若地下运输网络是一次旅程的一个终点，我们在另一个终点会发现什么？此时，同样地，由于城市不同，所得到的结果也有所不同。在纽约的洛克菲勒中心，我们发现了摩天楼——在其建立的时代（1930-1939），这是该城市最重要的建筑物。这里的上层建筑无法与基础设施相比，这是因为地铁的存在不是上层建筑存在的条件之一。

在蒙特利尔，因为地下空间的产生方式，城市中心的许多车站都与房地产交易有关——主要是办公室，这是市政当局所鼓励的。这里的机制是特殊的；在60年代，城市买下了对应未来地下线路的地皮，从而允许完成未来地下线路、控制进入并在这些关键点上促进城市开发。然后这些关键点被拿出来出售，包括进入地下的地役权和在私有土地和车站之间的公共空间建立通道的义务。因此，国家在通道的建设上省了钱，房地产开发商则将这些通道用作销售地面之上办公室的砝码。因此，上层建筑的发展是地下运输网络的直接结果，城市将其用作刺激城市发展的一种工具。

在东京，较大的地下空间意味着不同运输方式之间的转乘。因此，旅程的其他终点通常都是另一个车站或位于地下高速公路的端点的大型停车场。然而，在山手线非常重要的节点上出现了非常特殊的现象。伴随了二十年代城市扩张的私营企业在铁路领域之外也发展了商业战略，从而补偿其在铁路相关运营方面的盈利能力不足。这种多样化有三种形式：在线路的总站上建立花园城市区；在线路的农村终点建立休闲公园；和在城市一侧的终点站建立百货公司。

创造出将顾客转变成旅行者的需求后，以下步骤可确保忠诚的旅行者在离开该网络去另一家公司前尽可能地消费。这就是复合式建筑群的作用所在，该建筑集商店、餐厅、剧院和博物馆为一体，且各建筑都位于车站上方约10个楼层的位置，作为地下商业购物中心的补充。天价土地迫使开发商在所有方向都进行开发，从而产生不可比拟的功能性纠缠和空间复杂性。

在雷阿勒商场，没有摩天楼；甚至也没有任何房地产交易。在地面上仅仅只有几座楼阁。这是巴黎的中心；纪念物和博物馆、古城区和商业街区构成旅程的其他终点。正如我们所看到的一样，造成这种情况的原因与该地修建的历史有关。同样地，也与运营方式相关。这是国家决定、实现和资助的力量。尽管雷阿勒商场远非最大的城市地下经营（每日人流为100万人，而新宿区每日人流为500万人），但是其地下部分所占比例确实是最重要的。

地下空间的类型有哪些？

地下环境包括具体的细节，如界限、是否缺少自然光和有否障碍物，这与空间的组织密切相关。相对上层建筑而言，这是竖向反转：下降到地面以下，而非上升到地面以上。由于定向困难，这会导致安全和疏散等问题出现压力并复杂化。通过观察世界各地的一系列内部城市，我们可以看出各自的方法都与各国这些空间的舒适度相关。由于所使用的方法能决定公众和权威机构是否接受地下空间。

因此，就这点而言，需要指出直到六十年代地下空间的整体状况是非常令人失望的。第一代地下空间包含两种类型的站点：被视为改善现有路线的站点和表示只用于旅行的站点。我们发现蒙特利尔属于第一类站点。在-20℃温度和积雪一米深的情况下，尽管地下只有较低的天花板且类似一个迷宫，地下旅行相对地面旅行而言还是更受欢迎。通过空间轮廓和装修的多种多样可以看出大多数私营运营商没有任何规划或协调。动画和商业活动表明各分区之间存在巨大差异，且变化的营业时间使新手使用网络变得复杂化。直到2003年统一的传信系统代替本地定向和信息系统后，才使人认清形势。

尽管如此，蒙特利尔的内部城市取得了无可争辩的成绩：其他地下车站的品质。各个车站都是不同的，且都有艺术家的特色。且地下车站是通过中层楼与步行网络相连接的，从而形成壮观的空间情景并给人地下十字路口的感觉。

大多数私营运营商也意识到东京的地下网络也存在与蒙特利尔一样的缺陷。东京的地下网络属于第二类地下线路：用户无法选择。在大多数毗连区，旅行者会被困住。购物中心的概念化只需要考虑两个因素：流量和经济效率。但这两种第一代空间的相对或绝对束缚并未促使他们考虑进行空间创作和寻求舒适。

然而，随着七十年代地下世界的实现，演变变得非常明显。在蒙特利尔以及东京，出现了更多的宽敞的空间。最重要的是，通过玻璃嵌板和露台，空间拥有更多的自然光。这种演变在雷阿勒商场也是最而易见的，新的广场使空间具有10米高的楼底高度——几乎是传统广场的楼底高度的三倍。然而事实是地下空间的潜质依赖于结构因素。只有较大的空间才能允许进行空间组织使视野清晰，从而让地方变得更舒适、更宜人以及具有更少的压力。走廊的直线型布置是受到自然条件限制的。

单一围墙的布置更有可能创造充足的舒适空间。蒙特利尔最近某些开发项目就是这种情况，其中整部分都用于进行地下围墙建设。我们可以将蒙特利尔的内部城市视为围墙的延伸，通过位于道路下方的通道进行连接。雷阿勒商场的情况也是如此，地下空间安装在较大的外围围墙中。在大阪或新宿区情况相对有所不同，在大阪或新宿区，相当数量的购物中心位于道路上，因此受到公共容量的限制。尽管围栏区域有利于营造更充足的空间，但同时也会导致功能复杂化。由于其密度性，同样的'环境'范围内，雷阿勒商场各个方向的通道、网络和购物走廊相互缠绕。这些相同的通道、网络和大型购物中心倾向于以线性和连接的框架向外扩散。

地下空间营造会受到另一个非常重要的因素的影响：地下空间应符合防火规范。在法国，隔离是防范灾难的一个基本原则。在雷阿勒商场，这是通过将内部空间分成没有任何视觉或空间联系的不同部分而实现的。运输区域是一个封闭部分；新和旧广场是另外两个部分；地下通道又是另外一个部分等等。我们已经看到，在蒙特利尔，情况有所不同，风格为这个地方带来真正好的特性。在新宿区，站点地下自然采光是通过供车辆出入的圆形建筑实现的。此处的限制条件几乎都是无形的，且人们通常都不知道自己是否已离开这个车站。

地下城市的发展

除功能或空间差异或规则外，有一种特征可以让我们从根本上区别全世界各地的地下城市：城市发展的能力。洛克菲勒中心的地下购物中心几乎保持与战前一样没有任何改变，且自1986年开放新的广场后整个雷阿勒商场未进行任何修改。蒙特利尔和新宿区从未停止过扩展——前者扩展花了45年，后者扩展花了75年。这之间的差异是因为有一个单独的或多个决策者和介入者导致的。尽管洛克菲勒中心和雷阿勒商场是分阶段进行开发的，但它们都是全球运营型、建造时也是以此为目的的。约翰·D.·洛克菲勒考虑后决定、构想和资助自己的运营，就像国家资助法国的雷阿勒商场一样。蒙特利尔的情况有所不同，每个新的房地产交易可能由毗邻地区连接到现有网络，从而形成新的连接。在新宿区，各个新的线路都建造了自己的车站，从而与此前的车站进行连接，并通过在上层建筑中定期建立新的地下购物中心来进行发展。独立运营商的地块规模，从定义上看就只有最小限度的可扩展性。相反，多运营商系统中每一个新加入者都会把自己的区域变成网络的延伸。

同时，考虑可能引起地下城市发展的其他特殊因素也是非常有必要的：蒙特利尔的气候和不间断增长的流量和东京房地产价格的巨大压力。这些会刺激以盈利为目的的地下通路和商业中心的发展。反过来，法国建筑法规的演进对地下空间不利。接受公众的设施（ERP）再也不能够将人带到超过6米的深度——车站和停车场除外。最重要的是，未来政策将限制在地下发展商业。

这解释了属于巴黎公共交通公司（巴黎交通自主运营商）的走廊网络可继续在巴黎地下适度扩展，但主要是为旅行者的流通服务。建立自动人行道是非常重要的，可加速这些走廊的速度，这表明地下旅行是多么的无趣。同时也表明运输机是最重要的。

改建后的生活

在六十年代末，有一个得到广泛认同的观点是有必要重新考虑是否仍将雷阿勒商场用作商场。人们认为新的活动可能使这片地区恢复活力，这是公众电力公司计划的一部分。然而，在连续放弃法国政府部门、法国国际商业中心和大礼堂的计划后，人们对这个领域的未来逐渐开始产生疑虑，迁移至汉吉斯的批发商和贸易商使数以百计的建筑被遗弃，然而从八十年代开始人群开始返回，且该地区得以改造。

这次复兴是由四大因素造成的：大型车站，被视为吸引市郊人群的动力；广场，其位置决定其作为不同运输线路和地面之间的连接的催化剂；巴黎蓬皮杜艺术中心，地下城市另一个终点的再生器；和连接整个区域的宽阔步行街。这种方法使得建筑物逐渐被本质上为商业性的新活动所更新和取代，从而导致人口的巨大变革。随着不卫生的城市部分的消失以及土地成本的不断增加，苦的居民逐渐离开。

该区域的流动模式也有所改变；不断出入的批发商被步行者和游客等人流代替。区域中也出现完全不同的活动，每一天，每一刻都不相同。巴黎快速铁路网交通系统早高峰时段后，慢跑者和遛狗的人会出现，晚一点就是广场工人，他们会在花园享受午休时间。滚球球员、学生和老年人会在下午的时候占据主要通道，然后日光浴者会在草地上沐浴日光。刚学步的小孩的哥哥姐姐们会在放学后的下午加入他们。贩毒者也会在一个精微较远的距离闲逛。说唱和嘻哈乐队会在入口处演奏，期望获得

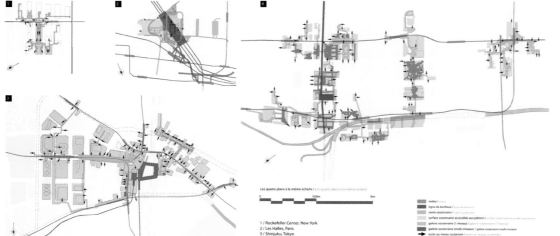

图1 Fig.1 同比例的4种地下交通系统图-纽约洛克菲勒中心、巴黎雷阿勒、东京新宿、蒙特利尔地下城 Four different underground systems drawn to the same scale – Rockefeller Center (New York), Les Halles (Paris), Shinjuku (Tokyo), and La Ville Intérieure (Montreal). © Seura Architectes Urbanistes

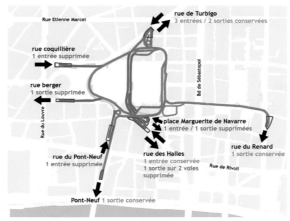

图2 Fig.2　环状地下公路演变 The transformation of the underground highway loop. © Seura Architectes Urbanistes

公众的关注。在一天结束的时候，步行者和情侣会在此欣赏日落余晖。

最近，在星期三下午和周末，可以看见许多来自市郊的青少年排列着来到快速铁路网交通系统。像人们在几十年以前在林荫大道上喜欢做的那样，这些青少年散着步，很快的融入步行街的人群中。他们对雷阿勒商场的概念是雷阿勒商场有一系列服装店，在这里他们很少买东西，但是会遇见很多女孩；同时也是快餐店和约会的地方，如纯洁喷泉。他们的线路开始于车站，相对新广场而言，更偏爱旧的广场，然后是昵称为"tube"勒斯高的通道。然后，通过行人街道形成一个大圈，越过花园，然后终点回在车站。他们在人群总能发现在住宅区所不存在的东西：可以匿名地游荡在这个多姿多彩的世界中（毫无困难地）。在这里，他们可以找到滑冰者、时髦人士、美国人、流行乐师、新潮人物、小流氓、古典学者、'borros'和其他人。在雷阿勒商场，来自市郊的青少年只是一个普通青少年。

雷阿勒商场的新组织

尽管从人流量的层面上说，雷阿勒商场算得上是个无可争议的成功，但在七八十年代人们对雷阿勒商场评论形形色色。批评家谈论这个地方的不舒适、空间的退化及可疑的美学。这个地方从未让巴黎人或住在法兰西岛的居民甚至游客心中产生过自豪之情。这个令人难以琢磨的地方分为地下和地上部分、车站和花园、商店和设施，在这座迷宫中，每个人有着自己的体验，此地的分割意味着不能描绘出统一的形象。缺乏了统一形象，对其运行带来很大不便，也对到其他地方乘坐交通工具增加了很大的难度。2002年以前，迎来了对此情况及进展可能性作出的反应。这座城市已决定在雷阿勒商场区内发起重塑过程。

计划开始的三十五年后，已无必要翻新了。肮脏的地方已消失，散乱的外观已被翻新。尽管对项目的思考包括整个地区和地下基础设施，实施干预的区域却被缩窄到由广场和花园组成的中心空间区域。与六七十年代大量工程相比，不同的是针对预可行性研究而展开竞争的四支队伍都考虑到了这个地方的深度。从这个地方的重塑从零开始（完全重建雷阿勒商场），每个队让自己置身于不同程度的思考。

在评选官员做出决定之前，四个工程少有地被呈现在一次展览中。此次展览的成功——十二万五千名参观者——展示了巴黎人和住在法兰西岛的居民的兴趣所在。这次展览过后，受到道不清说不明的对最终结果有影响力的情绪的鼓舞，讨论和辩论随之而来。因为没考虑到已深入到此地生活的基本变化、地下维度的细节，这些辩论看起来又在以前停滞不前的地方陷入僵局：认为这个项目是个建筑方面的大玩笑的想法再次拒绝"整合"要求，野心勃勃的工程不愿居民每天往来穿梭。

这些过分简单的方法——只用高度来表达自己的建筑；一个公共空间不是一项有雄心的工程——为了讲述一个都市建筑修建的复杂性说服我们写一本书，包括城市的地上和地下世界，一步一步地阐释已做的选择。因此，以上的这项工作是一个方法的全部阐明，解释着与本项旨在将雷阿勒商场重新整合进巴黎中心一系列大型公共空间的工程的环境、风险和目标。

屋顶变花园

巴黎的中心在哪？我们希望巴黎的中心是什么样子？这就是这场争辩所要回答的。这场巴黎与其他主要欧洲城市之间的对决能产生深刻的社会后果：巴黎正在经历的因大众旅游和低层住宅高档化而加速导致的人口减少和灵魂丧失。通过强调其自身独特的气质和在这大规模的巴黎人公共空间的庞大系统中重塑雷阿勒商场，巴黎必须也能够回应这个挑战。

为了实现居住、服务、零售和游客活动有机结合，应扩宽巴黎的中心，而不是允许一个独特的有特权的超中心存在。这有助于保持这个城市的社会多样性。我们都有雷阿勒商场临近一带的一系列纪念品，但我们也在这个地方有过支离破碎的经历。你可能是个巴黎人、一个郊区居民、一个本地人或一名游客、一个旅行者或一名路人，你可能会去游泳池、电影院、Fnac商店、巴黎电影影像中心、圣丹尼斯路、Agnès B或Freelance。许多人甚至不离开地下的区域快铁。每天有八十万人在广场穿梭，人与人之间的期望不同，有时甚至是相对立的。

雷阿勒商场地区有着相当多的有利条件。便利的交通（三条快速线和四条地铁线），公共活动和私人活动的混合，以及二十世纪七十年代重塑后留下的空闲区域（400 x 180米）。

它也经历了大量的机能障碍，一些人选择视而不见，一些人选择低估或高估它们。为了解决雷阿勒商场的问题，我们以渐进但基本的方式提出了一个针对改变区域形象的策略。

我们的建议

首先，通过压制或压缩地下道路的隧道入口，可能鼓励人们从林荫大道步行到塞纳河，从卢浮宫步行到波布尔（蓬皮杜艺术中心）[图2]。第二，为了给更新的贸易交易所（专注于时尚与文化）与城市、花园、更新的广场之间创造一个连接，对本地区进行重建。这个连接由一个横跨整个区域二十二米宽的商业街——类似"兰布拉大道"——而实现。在圣厄斯塔什教堂的边上，林立着一系列草坪和"绿色沙龙"（绿色屋子），在伯杰街的一边，林立着一系列运动场、花圃和报刊亭。购物中心创造了一个引人注目的10英亩大的花园，在此，不同年代的人对此使用的多样性得以实现的。[图3]

这个商业中心通过以前的广场（如今在一座九米高的屋顶下它完成了室内通道）连接勒斯高街。1830年以来，室内通道已为巴黎人熟知，但雷阿勒通道采用了现代技术。在我们的建议书中，再次诠释了宽阔楼式的巴尔塔设计，一个1,325 x 1,325米的屋顶覆盖了通道，一个从各方面来说都很奇特的屋顶，包括它的尺寸和由铜做成的开放"盒"设计。它比现在的Willerval阁楼低一些，与周围的树和谐地融为一体。

我们建议的屋顶大约两米厚，白天过滤光照，在夜间闪光。如果邻里和路人从贸易交易所、波布尔的全景露台和巴黎不同高度的地方观看它，它就像第五个水平立面。从内部看，它就像是一个新的"雷阿勒"广场。

与现在的勒斯高隧道形成对比的是，方格与其屋顶不同程度地为使用者和所有游客提供了一系列访问点和更好的能见度。它提供的屋顶代替了现在的窗户，维持了舒适的温度。从地平面能见到的广场更低处成了广场活跃的中心。广场边上的两种不同高度的地方有商店、有能提供服务的地方：在花园的一边是音乐学院和露台，在街道和人行道的一边是商店、警察局和各种各样的设备。

重建的区域快车站将变成一个阳台，朝向火车站台。吸收自然光、有巨大空间的火车站，能使游客轻易就找到路。安全问题将得到处理。将火车的运费用来供给运营倒是个办法。

在无止尽和损伤极大的革新后三十年，雷阿勒商场的未来再一次处在了紧要关头。我们已从过去的错误中吸取教训。我们建议公共区域的规模适当、长期存在的物料、邻近友好的施工地允许所有活动得以继续。雷阿勒商场能成为一个对每个人开放、受到每个人欢迎的主要友好公共空间。我们工程的规模能和位于巴黎人主要的地域[图4和图5]的其他物体高度共存，例如杜伊勒里宫、卢浮宫庭苑、巴黎皇家宫殿、波布尔和浮日广场等等。通道、花园、伸向花园的屋顶在保持不改变的同时允许巴黎的中心改变。

尾注：雷阿勒商场的建设现已起步。由Seura Architectes Urbanistes为重建而设计的广场、花园、道路、地下人行通道区域、由建筑师Berger Anziutti Architectes设计的华盖正在建设中[图6-10]。

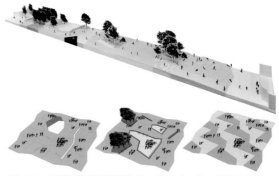

图3 Fig.3　算法模式花园的设计原则 Principles of the garden with an algorithmic pattern. © Seura Architectes Urbanistes

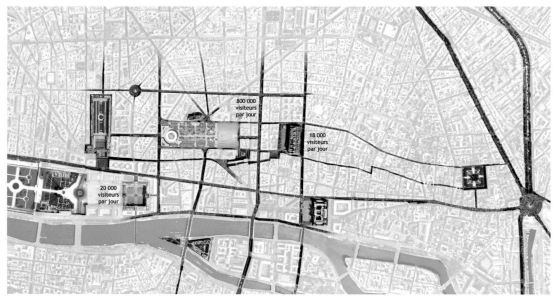

Fig.4 Les Halles in the system of the big Paris public spaces. © Seura Architectes Urbanistes

The Interior Cities

By the end of the 1960s, various cities worldwide had already experimented with underground urbanism, the principal ones being Montreal, Tokyo, Osaka, and on a smaller scale, New York [Fig. 1]. Each situation is different, but there is one common factor: the presence of an underground transport network, which appears to be the condition and the motor for the development of these underground spaces. It is this that generates the indispensable flows to make these places (that would not naturally attract the crowds) alive.

The underground space appears therefore firstly as the route of a journey. This route links to the surface via the underground station, or to another station underground. In Montreal, the access to the underground transport network is made from the lobby of a residential building. From there, one crosses an underground link into a commercial shopping centre. It can also be accessed from the neighbouring building, which joins underground to the basement of the first building. This pattern continues until the connection between two stations 750 metres apart can be made completely underground through corridors that join the successive undergrounds of the buildings of the city centre.

In Tokyo, the circular Yamanote line travels around the centre of the city. The connections between the suburban train lines, the metro network, and the Yamanote line are made through commercial shopping centres (for example, in Shinjuku), which absorb, with difficulty, the very heavy flows of pedestrian traffic. The dependence on the underground for the flows of transport is particularly obvious in Les Halles, Paris, where the relationship between commerce and movement has been totally reversed. The activity and the crowds already existed here, and the commerce generated the movement —until the saturation of the district. The commercial forum is now tributary to the transport, and 60 per cent of the visitors arrive through its underground stations. Therefore, any interruption in the underground transportation network results in immediate consequences that can be compared to those over ground.

As well as being the first place of voyage, the underground space can also be a destination. This is the case in Les Halles, where half of the visitors return to where they came from. A significant number of them arrive by the metro or the express RER network, and do not venture to the surface. This is equally the case in Montreal, where the interior city is more welcoming than the surface during the long Canadian winter months. It is, however, much less the case in Tokyo, where commerce has become an accompaniment to the underground corridors and is to some extent trivialised by this function.

What Types of Spaces for Infrastructures and Superstructures?

If the underground transport network is one extremity of a voyage, what do we find at the other extremity? Here, once again depending on the city, the results are diverse. In New York's Rockefeller Center, we find the skyscraper – the most important addition to the city in the era of its creation (between 1930 and 1939). The superstructure here is not comparable to the infrastructure, to the point where the presence of the metro is not a condition for the existence of the superstructure.

In Montreal, because of the way underground spaces have been produced, most stations in the city centre correspond to real estate transactions – essentially offices that have been promoted by the municipality. The mechanism here is particular; during the 1960s, the city bought the plots corresponding to the future underground lines to permit their realisation, control their access, and initiate urban development on these key points. These plots were then put on sale with the servitude of access to the underground and the constraint of an obligation to construct this access in public space, between the private land and the station. The State has therefore made an economy based on the realisation of the accesses, and the real estate developers of the buildings have used these accesses as an argument for the sale of their surface levels as offices. Thus here the development of the superstructures is a direct consequence of the underground transport network, using it as a tool of incitement for urban development by the city.

In Tokyo, the large underground spaces correspond to the moment of transfer between different modes of transport. Here, the other extremity of the voyage route is more often than not another station, or the large car park that ends the underground motorway. However, a particular phenomenon has developed on the most significant nodal points of the Yamanote belt. Private companies, which accompanied the urban extension into the suburbs from the 1920s, have developed commercial strategies outside the rail domain to compensate for the insufficient profitability of their rail-related operations. This diversification has taken three forms: the construction of garden-city zones on the main stations of the route; the realisation of leisure parks at the rural extremities of the line; and the creation of department stores at the final stations on the city side.

The following step was to assure that travellers consumed as much as possible before leaving the network for another company. This is the role of the giant complexes uniting shops, restaurants, theatres, and museums, which each company has built over approximately ten floors above its station, adding to the commercial shopping centres underground. The astronomical price of land here forces the operators to develop in all directions, leading to an incomparable functional entanglement and a spatial complexity.

At Les Halles, there are no skyscrapers; not even a real estate transaction. On the surface there are only several pavilions. It is the centre of Paris; monuments and museums, ancient districts, and a very commercial pedestrian sector constitute the other extremity of the voyage. The reasons for this, as we have seen, relate to the history of the redevelopment of the site. They are equally due to the type of operation. Here it is the force of the State that decides, realises, and finances. Even if Les Halles is far from the biggest urban underground operation (with one million people per day here compared to five million in Shinjuku), it is nonetheless the one where the underground part is proportionally the most important.

What Type of Space for the Underground?

Underground surroundings incorporate specific particularities, such as confinement, absence of natural light, and entanglement, which are related to the organisation of the spaces. There is an inversion of the vertical in comparison to superstructures; one descends below ground level instead of rising above it. This, along with the difficulty of orientation, can have a stressful effect and complicate the problems of security and evacuation. By looking at a range of interior cities worldwide, we can see the respective approaches concerning the comfort of these spaces in various countries. This approach taken is very important because it determines the acceptance of the underground place by both the public and the authorities.

At this point, it is necessary to state that an overall view of the sites realised up to the 1960s is disappointing. This first generation of underground spaces contains two categories of sites: those that are seen as an improvement to the existing route, and those that represent the only journey possible. Within the first category we find Montreal. At -20° C and a metre deep in snow, the journey underground is always preferable to the one on the surface, even if it has low ceilings and resembles a maze. The multitude of intervening private operators, without plan or coordination, is seen through the large diversity in spatial outlines and in decoration. The animation and the commercial activities show large discrepancies between sectors, and the varied opening hours complicate the use of the network for the novice. Concerning the way in which to orientate oneself within the space, it was not until 2003 that a unified signage system replaced the local systems of orientation and information.

There is nonetheless an incontestable achievement for the interior city of Montreal: the quality of its underground stations. Each one is different and features the intervention of artists. Their connection to the pedestrian network is made by mezzanines, which create spectacular spatial situations and give the impression of underground crossroads.

The underground network in Tokyo, also realised by a multitude of private operators, presents the same faults as that of Montreal. It comes into the second category of underground routes: those where the user has no choice. In most of the connection zones, the traveller is held captive. The conceptualisation of the shopping centres has only taken into account two factors: the flows and the commercial efficiency. But the relative or absolute captivity in these two first-generation categories of spaces has not pushed those conceiving them to create spatial inventions and search for comfort.

However, an evolution is apparent in the manner in which these underground worlds were realised during the 1970s. In Montreal, as in Tokyo, more ample spaces began to appear. More importantly, so did spaces featuring natural light by means of glass panels and patios. This evolution is also visible in Les Halles, where the new forum features spaces of ten metres in ceiling height – almost triple the height of those of the old forum. It is true, however, that the potential quality of the spaces under the ground depends on a structural factor. Only substantial volumes permit spatial organisation and thus clear views, which make the place comfortable, agreeable, and less stressful. A linear type of arrangement in the form of corridors is by nature restrained.

The arrangements realised within a unique enclosure are more likely to create ample and comfortable spaces. This is the case in certain recent developments in Montreal, where the entire sector has been used for the realisation of the underground enclosure. We can consider the interior city of Montreal like a succession of enclosures joined by passages underneath the roads. This is equally the case for Les Halles, where the underground spaces are installed in a large peripheral enclosure. This is less systematically the case in Osaka or Shinjuku, where a certain number of the shopping centres are situated on the roads and are thus limited by the public capacity. Although the enclosed area is favourable for the creation of ample spaces, it also leads to a complexity of functions. Due to its density, Les Halles presents a surprising entanglement of passages, networks, and shopping corridors moving in all directions within the same "surroundings" . These same passages, networks, and shopping malls have a tendency to spread out in a linear and connected framework.

Another very important factor influences the creation of underground spaces: they must comply to fire regulations. In France, isolation is a fundamental rule in the protection against disasters. At Les Halles, this materialises in the segmentation of the interior volumes into different parts without visual or spatial relationships. The transport areas are one closed part; the old and the new forums are two others; the underground passages are another, and so on. We have seen that the situation is different in Montreal and that the style brings a genuine quality to the site. In Shinjuku, the natural lighting underground to the site of the station is through the rotunda for car access. The limits here are mostly invisible, to the point where one often does not know whether one has left the station or not.

Growth of the Underground City

Beyond functional or spatial differences or rules, there is a characteristic that enables us to fundamentally distinguish underground cities worldwide: their aptitude for development. The underground shopping areas of the Rockefeller Center remain almost unchanged from their pre-war limits and the entirety of Les Halles has not been modified since the opening of the new forum in 1986. Montreal and Shinjuku have not ceased to extend – over a period of 45 years for the first, and 75 years for the second. The difference comes from having one individual or multiple decision makers and interveners. Even if they were developed in phases, the Rockefeller Center and Les Halles are global operations and considered as such. John D. Rockefeller decided upon, conceived, and financed his operation, just as the State did for Les Halles in France.

The situation is very different in Montreal, where each new real estate transaction has the possibility of linking through its neighbour to the existing network, and making a new link. In Shinjuku, each new line creates its own station in connection with those before it, and develops it by regularly installing new underground shopping centres and surfaces in the superstructure. The territory of a singularly run operation is by definition minimally extensible, whereas each new participant in a multiple operation makes its own part in the extension of the network.

It is also essential to consider some specific factors that lead to the growth of an underground city: the climate in Montreal and the incessantly increasing flows and the incredible pressure of the price of real estate in Tokyo. These incite the development of underground pathways and commercial centres for profitability. In the opposite way, the evolution of the construction regulations in France has made underground spaces unfavourable. Establishments receiving the public (ERPs) can no longer take people to a depth greater than six metres – with the exception of stations and parking structures. On top of this, future commercial surfaces to be installed underground will be limited.

This explains that the network of corridors belonging to the RATP (Autonomous Operator of Parisian Transports) can continue to spread – moderately – in the Parisian underground, but that it essentially caters to the circulation of travellers. The installation of moving walkways is significant, accelerating the speed of these corridors, and showing just how uninteresting the underground journey can be. It also shows that the transporter is all important.

图5 Fig.5 波布景观 General view looking to Beaubourg. © Seura Architectes Urbanistes

图6 Fig.6 兴建中的天棚 The canopy under construction. © Berger Anziutti Architectes (photo by David Mangin)

Life After the Redevelopment

At the end of the 1960s, there was a widely shared opinion about the necessity to rethink the use of Les Halles as a market. It was thought that a new activity would enable the rejuvenation of the district, as part of a plan scheduled by the public power. However, after the successive abandoning of plans for a Ministry of Finance, for a French Centre of International Commerce, and for a Grand Auditorium, doubts began to arise about the future of this sector. Wholesalers and traders moving to Rungis deserted hundreds of buildings, yet from the 1980s the crowds came back and the district was transformed.

This revival is due to four factors: the large station, which serves as an incontestable motor bringing crowds from the suburbs; the forum, which plays the role of catalyser in its position as a link between the different transport lines and the surface; the Pompidou Centre, regenerator of the other extremity of the site; and finally, the vast pedestrian area that unifies the entire district. This alchemy permitted the buildings to be progressively renovated and taken over by new, essentially commercial activities, leading to a drastic change in population. With the disappearance of the unsanitary urban sectors and the continual rise in the cost of the land, the poor inhabitants progressively left.

The pattern of frequentation to the district also changed; the incessant coming and going of the wholesalers was replaced by the influx of people such as walkers and tourists. Greatly differing practices began to appear on the site, succeeding each other depending upon the time of day and the day of the week. After the morning rush hour of the RER, joggers and dog walkers appeared on the site, followed a little later by the forum workers who would take their lunch breaks in the gardens. Boules players, students, and elderly people would occupy the main pathway during the afternoon, and then sunbathers installed themselves on the lawns. The toddlers profited from the parks before their older brothers and sisters came out of school to join them at the end of the afternoon. The drug dealers hung out at a slight distance. Rap and hip-hop bands would search for public attention at the entrances. At the end of the day, the walkers and lovers appreciated the sunset.

Recent times have seen – on Wednesday afternoons and weekends – youths from the suburbs arrive on the RER in tight rows. While strolling, as people used to on the Grand Boulevards several decades before, these youths blend without problem into the crowds that inhabit the pedestrian streets. Their notion of Les Halles is made up of a succession of clothes shops where they buy little but meet girls; fast food venues; and places to hang out, such as the Fontaine des Innocents. Their route begins at the station, favouring the old forum over the new one, followed by the passageway nicknamed the 'tube' Lescot. It then forms a big circuit through the animated pedestrian streets, neglecting the garden and finishing at the station. They find in the crowd that which does not exist in the housing estates: anonymity and the possibility of frequenting (without trouble) the world of all styles. There they find the skaters, the chics, the Americans, the goths, the trendies, the wide boys, the classics, and many others. At Les Halles, a youth from the suburbs is a youth among others.

The New Organisation of Les Halles

Though it is an incontestable success in terms of visitor numbers, a review of the changes to Les Halles in the 1970s and '80s is however quite mixed. The critics talk of the discomfort of the place, the degradation of the spaces, and questionable aesthetics. The site has never taken pride of place in the heart of Parisians or the inhabitants of the Île-de-France, and not even in the heart of tourists. At this elusive site, which includes an underground and an over ground, a station and a garden, shops and facilities – in this puzzle where everyone has their own practice – the division of the site means that no united image is portrayed. To this lack of unifying image are added significant problems in its functioning and in the accessibility of places for transport. By 2002, the time had therefore come for a reflection upon the situation and the possibilities of evolution. The city decided to relaunch a process of reinvigoration in the district of Les Halles.

Given that the unsanitary sectors have disappeared and the dishevelled facades have been renovated, and even though the reflection on the subject includes the whole district and the underground infrastructures, the domain of intervention has been narrowed to the area of the central space constituted by the forum and the garden. What is different from the numerous projects of the 1960s and '70s is that the four teams competing on the pre-feasibility study considered the depth of the site. Each of them 'positioned' themselves at different degrees on the scale, which goes from reorganisation of the site to the clean-slate approach.

In an extremely rare occurrence, the four projects were presented in an exhibition before the elected officials had made their choice. The exhibition's success – 125,000 visitors – shows the interest of Parisians and the inhabitants of the Île-de-France in Les Halles. Debate and controversy followed this occasion, encouraged by the vague sentiment of being able to influence the final decision. Without taking

图7 Fig.7 天棚内部立面原型 Prototype of the internal facades of the canopy. © Berger Anziutti Architectes (photo by David Mangin)

into account the fundamental evolutions, which have come into the life of the site, and the particularities of its underground dimension, these controversies seem to restart where they once stopped: the expectation of a large architectural gesture once again opposes the demand for 'integration' and ambitious projects oppose the daily practices of the inhabitants.

These simplistic approaches – architecture only expresses itself in height; a large public space is not an ambitious project – convinced us to write a book in order to detail the complexity of the process of an urban project, including the surface of the city and the underground world, and to enlighten, at each step, the choices that have been made. This work is therefore above all the entire explication of an approach, bringing light to the context, the risks, and the objectives that have supported this project that aims to reintegrate Les Halles in the series of large public spaces in the centre of Paris.

A Roof into a Garden

Where is the centre of Paris? What kind of centre do we want for Paris? This is what this contest is all about. The competition between Paris and other major European cities could have deep social consequences: a depopulation and a loss of soul accelerated by the processes of mass tourism and gentrification, which Paris is already experiencing. Paris must, and can, answer this challenge by emphasising its unique qualities and by reintegrating Les Halles within the larger system of large-scale Parisian public spaces.

The centre of Paris should be broadened in order to achieve the appropriate mix of residencies, services, retail stores, and tourist activities, rather than having a unique and privileged hypercentre. This should help maintain social diversity within the city. We all have an array of souvenirs of the neighbourhood of Les Halles, but we also have a very fragmented experience of the area. You might be a Parisian, a suburbanite, a local or a tourist, a traveller or a bystander; you might go to the pool, the movie theatre, the Fnac store, the Forum des Images, the rue Saint-Denis, Agnès B, or Freelance. Many people don't even get out of the underground RER station. Some 800,000 people go through the forum every day, each and every one with different and sometimes contradictory expectations.

The area of Les Halles has quite a few assets. It is very accessible (with three RER lines and four metro lines), a mix of public and private activities, and an empty space (400 x 180 metres) left by the restructuring of the 1970s. It has also experienced an important number of dysfunctions, which some people choose to ignore and others choose to underestimate or overestimate. In order to solve the problems of Les Halles, we propose a strategy aiming at changing the image of the area – in a progressive, yet fundamental way.

Our Proposal

Firstly, the possibility for everyone to walk from the boulevards to the Seine river, from the Louvre

图8 Fig.8　天棚棚顶原型Prototype of the roof of the canopy. © Berger Anziutti Architectes (photo by David Mangin)

to Beaubourg, is encouraged by suppressing or compacting the tunnel entrances for the underground roads [Fig. 2]. Secondly, a reorganisation of the area is undertaken so as to create a connection between a renovated Bourse du Commerce (dedicated to fashion and culture), the city, the garden, and a renovated forum. This connection is made by a 22-metre-wide mall – a ramblas [route] – crossing the whole area. This mall offers, on the Saint-Eustache church side, a series of lawns and salons de verdure [green rooms], and on the rue Berger side, a series of playgrounds, flowerbeds, and kiosks. The mall creates a highly visible ten-acre garden, where a diversity of uses is possible for different generations [Fig. 3].

The mall reaches rue Lescot through the former forum, where it becomes an indoor passage under a nine-meter-high roof. Since 1830, indoor passages have been familiar to Parisians, but the Halles passage uses modern techniques. Reinterpreting the famous Baltard design of wide pavilions, the passage of our proposal is covered by a 145 x 145-metre roof – an exceptional roof in all respects, including its size and its open 'box' design made of copper. Lower than the existing Willerval pavilions, it fits harmoniously with the surrounding trees.

Around two metres thick, our proposed roof filters light during the daytime, and glistens at night. In a way, it is like a fifth horizontal facade for the neighbours and bystanders viewing it from the vantage points of the panoramic terraces of the Bourse, Beaubourg, and various heights of Paris. From within the forum, it is a new 'carreau [square] des Halles'.

By contrast with the existing Lescot tunnel, the carreau and its roof offers a series of open access points at all levels, and better visibility for the users of the metro and all visitors. It also offers a roof replacing existing windows and preserving comfortable temperatures. The lower part of the forum, visible from ground level, becomes the lively heart of the forum. Two levels on the sides of the forum offer a mix of stores and services: conservatory and terraces on the garden side; and stores, police station, and various equipment on the street and passage side.

The restructuring of the RER train station will transform it into a balcony looking onto the train platforms. The station, a large space receiving natural light, will allow travellers to easily find their way. Security issues will be dealt with. Railroad freight to supply the forum is an option.

Thirty years after an endless and traumatising renovation process began, the future of Les Halles is again at stake. We have learned from the mistakes of the past. Our project proposes public spaces at the proper scale, long-standing materials, and a neighbour-friendly construction site allowing the continuation of all activities. Les Halles could become a major public space – popular, open to everyone, and friendly. The scale of our project is highly compatible with others at major Parisian sites [Fig. 4 and Fig. 5], such as the Tuileries, the Cour Carré du Louvre, the Palais Royal, Beaubourg, the Place des Vosges … The passage, the gardens, the 'roof into the garden' should allow the centre of Paris to change while remaining the same.

Endnote

Construction is currently underway at Les Halles. A design by Seura Architectes Urbanistes for the redevelopment of the forum, garden, roads, and underground pedestrian area of the halls is being constructed. A design by Berger Anziutti Architectes for the canopy is being constructed [Fig. 6 to Fig. 10].

图9 Fig.9　工程进展中,背景为波布,前景为在建的主体结构及黄色原型设计
A general view of the work in process. Beaubourg in the background, the structure under construction and the yellow prototype in the foreground. © Berger Anziutti Architectes (photo by David Mangin)

图10 Fig.10　工地场景A general view of the working site. © Berger Anziutti Architectes (photo by David Mangin)

主题IV：
案例分析2/4
Subject IV:
project discussions

亨克·贝克林
荷兰代尔夫特理工大学建筑学院
教授
杰士伯·尼杰维德
荷兰城市规划学者

Henco BEKKERING
Professor, Faculty of
Architecture
Delft University of Technology
Delft, Netherlands
Jasper NIJVELDT
Urbanist
Amsterdam, Netherlands

具有中国特色的乡镇的设计
Design of a Township with Chinese Characteristics

城市设计构成背景的一部分。在任何时候任何地方，可实现的规划若有不同，都是可比较的。将城市设计与一个地方的特色联系起来是实现特殊本土意义的一种方法。引入上述想法后，本文提出了在首届亚洲立体城市竞赛第一轮中入围二等奖的一个社区设计："墙"，同时也是第二作者毕业设计的一部分。该设计以现有景观为基础，且利用了该地点的农地地段。与竞赛报名的总体思路一致，墙建于地段线上，为各种各样的住宅类型建立建筑房屋用地。该设计参考了四周都围绕着墙壁的中式空间感。这种惊人的简单出发点就构成"具有中国特色的乡镇"的设计。

Any urban design is part of a context. At any given moment in time and at any place, realisable programmes are comparable if not the same. To relate an urban design to the characteristics of a place is the way to obtain a specific local meaning. After introducing these ideas, the paper presents the design for a neighbourhood within the second prize-winning scheme of the first round of the Vertical Cities Asia Competition: The Wall, part of the graduation project of the second author. The design is based on the existing landscape and uses the pattern of agricultural lots on the location. Walls are erected, consistent with the overall idea of the competition entry, on the plot lines to create building lots for a variety of dwelling typologies. The design refers to the Chinese perception of space as enclosed by walls. The surprisingly simple point of departure leads to a design for a 'Township with Chinese Characteristics'.

背景

城市设计构成背景的一部分——一个更大的整体。即使设计在实现后，更多的是保持原样而非改变，而保持不变的才是占主导地位的。这仅仅只是关于尺度的问题：背景的尺度。插入一个事实就是，根据定义，可实现的规划在任何时候和任何地方都变得越来越相似，通过全球化力量加强的一个过程，让某一地方具有个性或被认同所必需的差异只可能来源于背景的特殊地方特色、活力和力量。城市设计不是一种自由的艺术形式。因此，作者确信一个有责任的城市设计回应并成为背景的一部分是非常必要的。这能够并且应该扩展到一个项目的实际背景，包括地方气候和文化、客户目标、政治敏感性和市民、居民和使用者的意见等方面。

这种广泛的背景只能从其历史进行理解，以理解为目的进行分析，而不是一定要以维护、恢复或修复为目的。通过参阅背景，设计师可从当前形势中提炼出其意义。从任何城市中得出的结果都是分层次的：城市的记忆的显示。历史层的易读性导致历史延续性。正如理查德·塞尼特所描绘的[①]，"城市环境中的'发展'相对于简单取代之前已存在的东西而言，是一个更复杂的现象；发展要求过去和现在进行对话，它是关于进化而非消除的问题。"现在我们认为城市是较大网络、甚至是其内部网络的一部分——网络由节点和节点之间的链接构成。若所有节点都有相同的特性，这些链接就没有理由存在了。这说明网络中节点的身份需要有所不同——公众即城市使用者所认可的身份：居民和游客。

记忆危机

整个二十世纪现代运动所产生的令人不安的结果之一就是大量现有的且通常具有历史意义的城市结构和建筑的拆毁。如果没有爆破技术，建筑师及其客户可能会自己拆毁建筑物。在西方国家，历史经验告诉我们，总会有那么一个时刻，"人们"不再接受在现代化华丽词藻掩饰之下的拆迁。他们会奋起游行表示反抗。

这些抗议的背后的心理事实是，人们对让他们觉得像在家中般轻松自在的熟悉的日常环境具有归属感：一个基本的人类防御机制。

此外，人们只有将自己的生活环境与其它环境进行比较，才能意识到自己的身份：他们自己的身份和一个地方的身份。这会形成一种具有社会文化意义的归属感：

归属于共享同一环境和感受的一个地方和一个社区。若一切进展顺利，就会对这些环境的情况形成一种共有的责任感。

若忽略这一点同时破坏城市环境的基本部分，在某些时刻可能会导致社会凝聚力的丧失和社会动乱。这样产生的集体记忆缺失被克里斯了·波伊尔形容为"记忆危机"[2]，现在出现于欧洲和美国，成为对社会凝聚力的一个严重威胁。它同时也可能导致地方认同的丧失，如今我们意识到这同时也会产生负面的经济效应，因为这会降低某些特殊城市特殊的吸引力，城市甚至会变得越来越普通。由此看来，这可能是目前快速发展中的国家在极快的城市发展之下面临的越来越大的陷阱。

历史延续性

城市规划中历史延续性这一概念于二十世纪80年代初期出现于欧洲和美国。该概念承认现有环境是过去景观、居住点和建筑物的发展结果，过去的环境不仅对过去特定社区的人有意义，对现在的人也有意义（尽管意义未必相同），其重要性在于它是社会生活条件社区建设的主要机制之一。此外，历史延续性承认过去和现在涉及城市设计的的形式方面——城市设计的理论和设计之间的关系。

尽管有人不愿意接受，但每个人类都是他或她所不能摆脱或否定的特定的文化环境中的知识和信息的累积。这些知识来源于日常经验、教育和培训。这个文化机制可证明设计方式选择的合理性，它能形成贯穿过去、现在和未来的不间断的主线。如果做得够好，过去、现在和将来将会被赋予对人们有意义的新的形式连贯性。为了保持他们的长期性，城市需要健全的技术物质和网络。同样重要的是，他们的物理结构包含某些意义，可作为使人们从精神上与其环境相连的方式，从而帮助保持社会和物理意义上的永久性。

参考历史延续性进行设计的选择同时也存在有效性——且不止只有一种。第一，如前所述，在处理历史环境时，它利用了人们对待历史环境时形成的情感纽带，从而可防止不必要的破坏，同时因其可为居民提供与一个地方联系并保持联系的机会能帮助建立和保持社区感。这可以帮助建立一种对环境的共同责任感，抵消忽视和破坏的趋势。因此，人们对某个城市的归属感通常至少有一部分是以其历史为基础的，哪怕该地点历史相对较短。

第二，这种设计方法能帮助仔细调整物理变化，从而有利于避免地上和地下建筑物和基础设施中累积的资本投资的不必要的湮没。

第三，（如上所述）由于根据定义可实现的规划在任何时候任何地点都是相似的，所以使一个地方具有自己的个性所需要的差异只能来源于其背景的具体特性。

具有中国特色的乡镇

自中国于1978年正式实行以市场为导向的经济政策后，城市规划水平和GDP得到极大增长。因此，城市面临着一场史无前例的大规模的超常速度的激进变革。中国城市已逐渐适应苛刻的市场力量，形成永久的变化状态。这状态可以作为提高生活标准的实用的调速轮。公寓住宅和基本福利设施可快速和高效成本地建立。

一些作者承认由于这次激进的变革，现代混合社会将要来临，带来更多的流动性和个人选择自由，并导致身份与地方无太大关联的观点[3]。

相反地，其他几位作者辩称，激进的城市主义和变革不仅仅改变了环境，同时也改变了人类与环境之间的关系和相互作用。陈飞、Ombretta Romice[4]和缪朴对其如何造成普通公共场所质量和使用的降低的问题进行了解释，且在潜意识层面上解释其如何导致城市逐渐产生"无所凭依"的感觉。Jung[6]断定"中国的资本化式现代化会导致人日常生活中真正意义的丧失，从而使人们远离其社会和自然环境。"中国城市庭院、街心花园和步行街道等的精细的网络快速地被现代的城市景观代替，这样的景观具有不同的行车路线、松散定义的空地、独立塔和私有化围地。

朱小地[7]辩解称延续性正面临着压力："建筑物是否与城市结构很好的融合在一起取决于其对公共空间的影响。你只有在将城市背景作为设计过程的出发点时，才能明确的定义一个建筑物的个体位置，然后成功获得一个较好的设计解决方案。现在，我们经常能够看到具有过度的个人风格的建筑物倾向于展示一些非常特殊的东西，从而试图表明其超过城市语境。这种行为会扰乱建筑物和城市环境之间的关系，同时也会毁坏其自身的特性。"更准确地说，公共空间起着越来越重要的分离作用，但因其只作为城市的基本元素而受到忽视。

人们对一个地方的使用和评价都受到其对空间的感知影响，本文所提出的城市设计项目正是基于这种观点[8]。因此，空间的感性经验在任何城市设计的成功中都是关键问题。而由于比西方人多得多的中国人在传统上倾向于以感性和直观方式来看待世界[9]，这一点就显得更为重要。因此，为了知道如何构建具有中国特色的全新社区，先了解中国人的空间知觉是非常重要的。标题为"墙"的项目旨在为近代的城市规划提供一个选择，它更注重景观和物体，而非空间本身。

空间知觉

为了了解中国人的空间知觉，应首先了解一种文化是如何理解和表示空间和公共性这两个概念。几个世纪以来，中国在数个世纪里累积演变而非彻底革命的过程中逐渐发展出其空间知觉。在最近的现代化之前，中国城市被看作一个整体，且通常是以适用于现有建筑类型的计划为基础的。这是一种集体艺术品。一些系统使用地原则是根据很早以前的先例而建立的。

这些经过几个世纪形成了空间知觉，但在当今的中国城市中却面临着巨大的压力。根据缪朴[10]以及李晓东和杨茳善[11]的研究，共有5个重要的知觉原则：线性、层级、统一性、人文尺度和外围。下文将对这些原则进行简述。这个城市设计项目是在地方背景范围内对上述原则的当代诠释。

线性。与公共生活相关的中心性在西方和在中国的认知有所不同。在西方，较大的中央和静态节点在公共生活中起着非常重要的作用，而在中国，较小的分散的公共场所和线性街道却是非常重要的[图1]。"上述空间知觉多半是对向前或向后运动的知觉"正如缪朴[12]

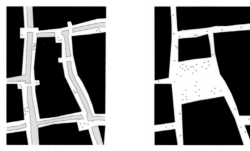

图1 Fig.1　左：中国线性公共空间 右：西式中央广场（根据缪朴1990绘）
Left: Chinese linear public space. Right: Western central square (after Miao 1990).

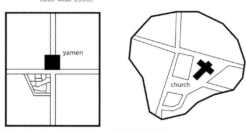

图2 Fig.2　左：中式层级组织 右：西式非正交城市肌理
Left: Chinese hierarchical organisation. Right: Western non-orthogonal urban tissue (after Miao 1990).

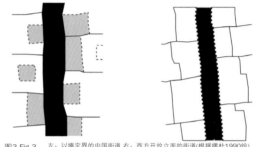

图3 Fig.3　左：以墙定界的中国街道 右：西方开放立面的街道（根据缪朴1990绘）
Left: Walls defined street in China. Right: Open façade to street in the West (after Miao 1990).

图4 Fig.4　上：中式平均分布的小尺度开放空间 下：西式中心街区的大尺度开放空间
Above: Evenly distributed small scale open spaces in China. Below: Large open space in centre of block in the West (after Miao 1990).

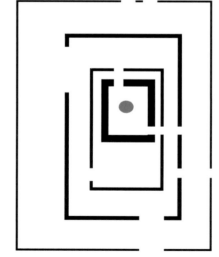

图5 Fig.5　一系列闭合空间：成都茅舍
Series of enclosed worlds: cottage, Chengdu.

所写。静态公共活动以不同的方式进行，如沿着街道的较小的空间。

层级。传统中国城市是按照层次化方式进行组织的[图2]。这种层级是一种成文法的结果，该成文法详细说明了理想城市的一系列规则。这反映出符合儒家思想的一种'美好'社会。第一眼看上去似乎是混乱的，但实际上这是一个很有组织性的分级系统。因存在上述这些极为严格的规定,所以不可避免地,各个城镇的认同只能通过有限的方法进行建立:现有的自然类型。这意味着中国的每个城镇都具有相似之处,但由于各自的类型不同,各个城镇也具有不同的空间结果。

统一性。中国文化深受儒家思想的影响。儒家哲学非常重视家庭和亲属关系,认为这是社会的基础。这表明,其介入范围下至个人传统房屋布局水平。中国家庭的私生活从传统意义上讲就像微观世界,通过墙围绕、保护和界定房屋。同时这也对与街道的关系产生了影响。通过私有空地将住宅主要房间与街道进行分离[图3]。导致的结果是(几乎还空白)墙可以界定街道。

人文尺度。墙形成的街道会影响城市的规模和比例,这从传统城市形态中可以看出。小型的开阔场地、庭院、花园和其它形式的空地在尺度宜人的城市中均匀分布,建筑物几乎都不超过两层[图4]。

围墙。线性、层级、统一性和人文尺度等不同原则可以与重要原则围墙相结合。空间的封闭把握住了中国空间知觉的本质[图5]。从根本上讲,空间被认为是一系列封闭的世界,较小的单元在缩小的规模中重复较大单元的形式。汉语中用空间一词表示"通过三维式的'封闭'或界定对空洞体积进行创造和排序"[13]。因此,空间可通过各种各样的围墙和不同空间序列的交叉形成;空间是动态的,可从非常公开的地方一直到私人卧室。因此,空间是一点点呈现的。下一个空间通常是不可预计的,从而会形成一种神秘感[图6和图7]。

在成都的设计
地方背景

城市设计的第一步是根据现有景观、基础设施和现场的建筑物建立一个框架。成都亚洲垂直城市比赛第一轮现场的景观呈阶梯状,容纳种植大米、小麦、蔬菜、茶树、草药、烟草、蚕丝和生产牛肉和猪肉的农地。场地的地势高度差很大[图8]。场地中有一个中央山谷,山谷中粮食具有很高产量。设计框架的第一关节就受到地势的直接影响。该山谷成为周围环境的郁郁葱葱的主心骨。现场其他有趣的特征就是竹林丛生的丘陵以及蓄水的池塘。可能的情况下应把这些全包含在内[图9]。

框架的第二关节是由现有道路形成的。沿这些道路的建筑物能与新的街区相结合。这些道路将在新的社区中的公共空间层级中形成干线。他们具有对称的剖面,且私人和公共空间具有明显的差异。连续立面、无台阶、(半)公共基层和最高六层的设计且具有混合的城市职能预计会带来人流和物流。该框架在现有情形下建立,下一设计步骤将会对上述五个知觉原则进行解释。

线性

下一个介入因素是在东西方向添加新的车道[图10]。这些车道沿着梯田和稻田的边缘。朝南的街道可充分利用冬季的(较稀缺的)阳光和夏季的盛行风。这些街道可容纳日常生活的流量。在公共和私人空间之间有一条长为5米的过渡地带,包含走廊、阳台和前院。街道的非对称剖面可为坐下歇脚、种植和路边售货摊提供空间。为了提高街道和建筑物之间的相互作用,需要每隔十米就

设置一个入口。基层应包含商店、餐厅、茶楼或其他半公共设施。建筑街区是可辨认、可控实体。

层级

得出的结果就是从公共空间到私人空间的分级系统。层级可通过车道的宽度的改变进行强调。本框架界定了建筑物区域[图11]。不同的规划都增强了分级,山谷四周具有混合的生活和娱乐规划。主要干线是商业、办公楼和住宅的混合区域,而二级干线主要的是住宅。从战略角度来讲,当景观、主要干线和地铁站聚在一起时,特殊的建筑能形成焦点元素。

统一性

为了将建筑区域发展成为城市街区,集体主义的想法占了领先地位。习惯上,中国家庭的房屋都建有墙,从而围

图6 Fig.6　成都文珠坊
Wenzhufang, Chengdu

图7 Fig.7　杜甫草堂竹径
Bamboo path, Dufu

绕、保护和界定住宅及其独特的部分。同时,该原则可扩大到集体住宅形式[图12]。各街区需提供共享和私人空间。与西方封闭的街区和独立塔相比,中国街区的私人空间分布更为均匀。

人文尺度

通过沿着现有圈地线建立2.7米高的墙[图13],可根据人文尺度创建私人空间,同时为各种各样的住宅类型建立建筑用地。在较小的圈地上,最多60%的建筑区域允许建筑高度超过四层。每所房屋都有一块敞开的私人空间。某些圈地是留作半公共空间和通道的。这些圈地上的墙应留有孔洞,从而允许穿过这些街区。独立单元的入口来自于这些共享空间。

结果:外围

最后的设计[图14]是"具有中国特色的乡镇"。它是通过位置的现有特征形成的框架,同时对中国文化基本的知觉原则进行解释。这提供一系列封闭的世界,具有规模宜人的庭院、花园和较小的开放场地。墙是最突出的物理表现。最终设计从俯视角度来看是非常混乱的[图15和图16],但从视线高度来看,显示的却是穿过一系列封闭世界的可以理解的运动[图17和图18]。

当代的诠释

本文提出的城市设计项目旨在对中国城市传统的一些主要原则进行当代的诠释。该项目表明,他们形成城市的持久元素,能不断吸收新的功能和模式。以墙为标志的外围形成了这其中的主要元素,从而一点一点的展示空间的神秘感。这通过建立和排序空体积把握住了中国空间知觉的核心。通过建筑元素的封闭,空间成为对日常生活非常有意义的特殊的'地方',能提供个性和集体地方认同。

Context

Any urban design is part of a context – a larger whole. Even after realisation of an urban design scheme, much more stays the way it is than changes, and what stays the same is dominant. It is a simple question of scale; the scale of the context. Adding the fact that by definition, realisable programmes are becoming ever more alike at any given moment in time and in any place, a process strengthened by the forces of globalisation, the necessary differences that give a place its individual character or identity can only be derived from the specific local characteristics, dynamics, and forces of the context.

Urban design is not a free art form. Therefore it is the authors' conviction that it is essential for a responsible urban design to respond to and become part of its context. This can and should be extended beyond the physical context of a project to include aspects such as the local climate and culture, the objectives of the client, political sensibilities, and the opinions of citizens, inhabitants, and users.

This broad context can only be understood from its history, which has to be analysed with the specific purpose of understanding it, though not necessarily maintaining, retrieving, or restoring it. By referring to the context, the designer distills meaning from the existing situation. The result in any city is layered: the display of the memory of the city. The legibility of the historical layers results in historical continuity. As Richard Sennett phrases it, "'Growth' in an urban environment is a more complicated phenomenon than simple replacement of what existed before; growth requires a dialogue between past and present, it is a matter of evolution rather than erasure."

We now think of cities as parts of larger networks and even as networks within themselves –networks consisting of nodes and links between those nodes. If all nodes had the same qualities, there would be no reason for the links. This implies the need for differences in identities of the nodes in networks –identities that are recognisable for the general public, the users of the city: inhabitants and visitors.

The Memory Crisis

One of the troubling results of the modern movement's dominance throughout the twentieth century is the breaking down of a great many existing, most often historical urban structures and buildings. If there had been no war-time bombing, the architects and their clients would break down buildings themselves. In Western countries, our experience is that there will be a moment when 'the people' will not accept the ongoing demolition under the rhetoric of modernisation anymore. They will start to rise and demonstrate against it.

Behind such protests is the psychological fact that people are attached to their familiar daily environment. It makes them feel at home: a basic human defense mechanism. Moreover, people can only become aware of their identity if they can contrast it to another: their own identity and the identity of a place. This results in a socially and culturally meaningful feeling of belonging: belonging to a place and to a community that shares its surroundings and that feeling. If all goes well, with this comes a shared feeling of responsibility for the state of those surroundings.

Neglecting this and breaking down essential parts of the urban environment will at some moment lead to the loss of social coherence and to social unrest. The resulting loss of collective memory, which M. Christine Boyer characterises as the 'memory crisis', is nowadays seen in Europe and even in America as a serious threat to coherence in society. It also leads to a loss of local identity, which we now realise has negative economic effects as well, as it reduces the specific attractiveness of any particular city. Each becomes ever more generic. It seems this might be the trap that is opening ever wider for the extremely fast urban development going on in rapidly developing countries now.

Historical Continuity

The concept of historical continuity in urbanism came up in Europe and the United States of America in the early 1980s. It acknowledges that the existing environment, as the physical result of developments in the past in landscapes, settlements, and buildings, had meaning for the people of a specific community in the past and still has meaning for people now (though not necessarily the same meaning). Its importance is in being one of the main mechanisms of community building as a condition for social living. In addition to this, historical continuity acknowledges the relationship between theories and designs from the past and the present that deal with the formal aspects of urbanism – with urban design.

Even when one is unwilling to accept this, in each human being there is an accumulation of knowledge and information within a given cultural context that he or she cannot shed or deny. This knowledge results from daily experience, education, and training. Awareness of this cultural mechanism justifies the choice for a way of designing that results in an unbroken chain from the past through the present to the future. If done well, past, present, and future are given a new formal coherence that makes sense to people. To retain their longue durée [long duration], cities need sound technical substance and networks. Just as important is that their physical

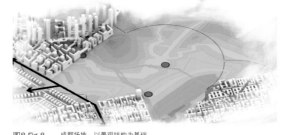

图8 Fig.8 成都场地：以景观结构为基础
Chengdu site. Landscape structure as basis.

图10 Fig.10 加入东西街道的基础设施网路
Infrastructural network with new east-west lanes.

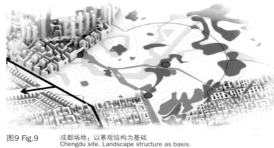

图9 Fig.9 成都场地：以景观结构为基础
Chengdu site. Landscape structure as basis.

图11 Fig.11 建设区
Building zones.

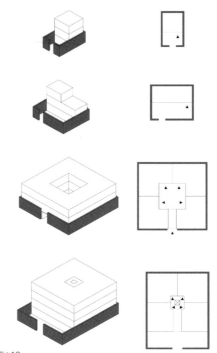

图12 Fig.12
不同住宅形式
Different housing typologies.

structure contains meaning as a way to enable people to connect –mentally –to their environment and thus to help sustain permanence, in a social and physical sense. The choice to design with reference to historical continuity also has a form of efficiency in it –even more than one. First, as stated earlier, it makes use of the emotional bonds that people develop when dealing with historical environments. Unnecessary breaks are prevented and this helps to establish and maintain a sense of community as it offers residents the chance to connect and to stay connected to a place. It helps create a common feeling of shared responsibility for the environment, counteracting tendencies of neglect and vandalism. Thus the attachment of people to an urban place is always at least partly based on –again –its history, even when it is a relatively short one.

Second, this way of designing facilitates a careful phasing of physical changes that helps prevent unnecessary annihilation of capital investments accumulated in buildings and in infrastructures above and under the ground. Third, since (as stated before) by definition, realisable programmes are alike at any moment in time and at any place, the differences necessary to give a place its individual character can only be derived from the specific qualities of the context.

Township with Chinese Characteristics

Since China formally adopted market-oriented economic policies in 1978, the levels of urbanisation and GDP have increased enormously. As a result, cities have been confronted with radical transformation on an unprecedented scale and at an extraordinary rate. The Chinese city has become hyper-adaptable to the demanding market forces, resulting in a permanent state of change. It works as a pragmatic flywheel for improving living standards. Apartments and basic amenities can be built fast and cost-effectively. Some authors acknowledge that due to this radical transformation, a modern hybrid society is rising enabling mobility and individual freedom of choice, and resulting in the idea that identities and places are more loosely related .

To the contrary, several other authors argue that radical urbanisation and transformation has not only changed the environment, but also the relationships and interactions of people with it. Fei Chen and Ombretta Romice and Pu Miao explain how it leads to a decline in the quality and use of ordinary public places, and on a subconscious level to a feeling that cities are becoming "placeless". Jung diagnoses that "the capitalized modernization of China leads to a loss of authentic meaning in the characters' daily lives, thus distancing them from their social and natural environments." The Chinese city's finely meshed networks of courtyards, pocket parks, and pedestrian-friendly streets are rapidly being replaced by a modern urban landscape with distinct travel ways, loosely defined open spaces, freestanding towers, and privatised compounds.

Continuity is under pressure argues ZhuXiaodi: "Whether a building is well integrated in the urban fabric is dependent on its impact on public space. You can only define the individual position of a building clearly and succeed in getting a good design solution when the urban context is the starting point of the design process. Today we very often see buildings which, being supported by an excessive individual style, tend to show something very special and thus try to rise up above the urban context. This behavior disturbs the relations between the buildings and the urban environment and it destroys its quality." More precisely, public space increasingly plays a separating role, and is ignored as being a basic element of the city.

The urban design project presented in this paper is based on the idea that the way people use and value places is influenced by their perception of space. Therefore, the perceptual experience of space represents a key issue in the success of any urban design. This is even more important since Chinese people, more than Western people, traditionally tend to see the world in a perceptual and intuitive way. It is therefore crucial to understand the Chinese perception of space in order to know how to structure it for a new neighbourhood with Chinese characteristics. The project titled 'The Wall' aims to provide an alternative to the recent urbanisation, which is more focused on the spectacle and the object than on space itself.

Perception of space

In order to understand the Chinese perception of space, one would have to understand how a culture perceives and formulates the ideas of space and publicness. China progressively developed its perception of space for centuries in aprocess of accumulated evolution rather than outright revolution. Before the recent modernisation, Chinese cities were conceived as a whole, and were usually based on a plan that was consistently applied to the existing topography. It was a collective work of art. A few principles were systematically applied following precedents established long before. These shaped the perception of space for centuries, but

图13a Fig.13a 墙的兴建 Erecting walls

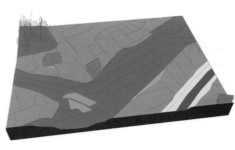

图13b Fig.13b 墙的兴建 Erecting walls

图13c Fig.13c 墙的兴建 Erecting walls

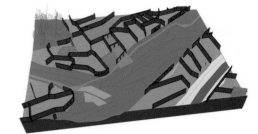

图13d Fig.13d 墙的兴建 Erecting walls

are under high pressure in the modern Chinese city. Based on studies by Miao and Liand Yeo, there are five crucial perceptual principles: Linearity, Hierarchy, Unity, Human Scale, and Enclosure. These principles are brieflydiscussed below. The aim of the urban design project is a contemporary interpretation of these principles within the local context.

•Linearity. Centrality in relation to public life is per ceived differently in the West than in China. In the West, large central and static nodes play an important role in public life, while in China small, scattered public places and linear streets are crucial[Fig. 1]. "The perception of these spaces is mostly that of forward or backward motion," wrote Miao. Static public activities take place in a different pattern, such as small spaces along the street.

•Hierarchy.Traditional Chinese cities were hierarchically organised[Fig. 2]. This hierarchy was the result of a written code that specified a set of rules for the ideal city. This reflecteda "good" society according to Confucian doctrine. What at first sight seemed to be chaos was in fact a very organised hierarchical system. Since there were such strict regulations, inevitably, the identity of each town could only be created with limited means: the existing natural topography. This meant that every town in China had a similar basis, but a different spatial outcome according to its topography.

•Unity. Chinese culture has beendeeply influenced by Confucianism. Confucian philosophy valuesfamily and kinship as the base of society. This shows, down to the level of the layout of the individual traditional dwelling. The private life of the Chinese family is traditionally like a microcosmos, with walls serving to enclose, protect, and define the house. This also influences the relationship to the street. A form of private open space separated the main room of a house from the street[Fig. 3]. The result was that(nearly blank) walls defined the streets.

•Human Scale.The streets,formed by walls, influenced the scale and proportion of the city, which can be seen in the traditional urban morphology. Small open areas, courtyards, gardens, and other forms of open space are distributed evenly throughout a human-scaled city with buildings rarely exceeding two storeys[Fig. 4].

•Enclosure. The different principles of Linearity, Hierarchy, Unity, and Human Scale can be synthesised with the key principle of Enclosure. The enclosing of spaces touches the essence of the Chinese perception of space[Fig. 5]. Space is fundamentally perceived like a series of enclosed worlds, and the smaller units repeat the forms of the larger one on a reduced scale. The Chinese word for space itself, kongjian, represents "the creation and ordering of empty volumes as a result of the 'enclosure' or bounding of three-dimensional elements." Space is therefore experienced through the crossing of various enclosures and different spatial sequences; it is dynamic, from the very public all the way to the private bedroom. Space is thus presented little by little. The next space is always unpredictable, which therefore creates a sense of mystery[Fig. 6and Fig. 7].

Design in Chengdu
Local context

The first step of the urban design is to set up a framework based on the existing landscape, infrastructure, and buildings on the site. The landscape on the site for the first round of the Vertical Cities Asia competition in Chengdu is terraced to accommodate agricultural lots that produce rice, wheat, vegetables, tea, medicinal herbs, tobacco, silk, beef, and pork. The

图14 Fig.14 结果 Result

site has significant differences in height[Fig. 8]. There is a central valley with a high production of grain. The first articulation of the design framework is influenced directly by the topography. The valley becomes a new lush green backbone for the neighbourhood. Other interesting features of the site are the hills on which bamboo thickets grow, and the ponds that store water. These are incorporated if possible[Fig. 9].

The second articulation of the framework is formed by the existing roads. The buildings along these roads can be integrated into new blocks. These roads will form the primary arteries in the hierarchy of public spaces in the new neighbourhood. They have a symmetrical profile with a clear distinction between private and public spaces. Continuous facades, no setbacks, (semi-)public ground floors, and a maximum of six storeys with mixed urban functions are expected to bring about a flow of people and goods. The framework is built upon the existing situation. The next design steps will interpret the five perceptual principles mentioned above.

Linearity

The next intervention is to add new lanes in an east-west direction[Fig. 10]. These lanes follow the edges of the terraces and rice paddies. These streets, open to the south, will take full advantage of the (scarce) sunshine in winter and of the prevailing winds in summer. The streets will accommodate the flows of daily life. There is a transition zone between public and private of five metres, which can contain porches, verandas, and front yards. The asymmetrical profile of the street offers spaces to sit, for planting, and for street vendors' stalls. To improve the interaction between the street and the buildings there needs to be an entrance every ten metres. Ground floors contain shops, restaurants, teahouses, or other semi-public facilities. The building blocks are recognisable and controllable entities.

Hierarchy

What results is a hierarchical system leading from the public to the very private. The hierarchy is emphasised by varying the width of the lanes. This framework defines zones for building[Fig. 11]. Different programmes strengthen this hierarchy, with a mixed programme of living and leisure around the valley. The main arteries attract a mixture of commerce, offices, and dwellings, while the secondary arteries will have mainly dwellings. At strategic points, where landscape, main arteries, and metro stations come together, special buildings can form focal elements.

图15 Fig.15　鸟瞰图 Birds' eye perspective

Unity

To develop the building zones into urban blocks, the idea of collectivity leads. Traditionally, the house of a Chinese family was built up with walls serving to enclose, protect, and define the dwelling and its distinctive parts. This principle has also been enlarged to collective housing forms [Fig. 12]. The blocks need to provide communal and private spaces. In contrast to Western closed blocks and freestanding towers, private spaces are distributed evenly throughout the block.

Human scale

By erecting 2.7-metre-high walls along the existing plot lines[Fig. 13], there emerges the possibility of creating these private spaces on a human scale, and creating building plots for a variety of dwelling typologies. On the smaller plots, a maximum of 60 per cent of the building area is allowed to a maximum height of four storeys. Every house enjoys thus an open private space. Certain plots are reserved for semi-public spaces and access. Walls on these plots will be cut with holes to allow penetration into the block. Entrances of the individual units are from these communal spaces.

Result: Enclosure

The design that results [Fig. 14] is a 'Township with

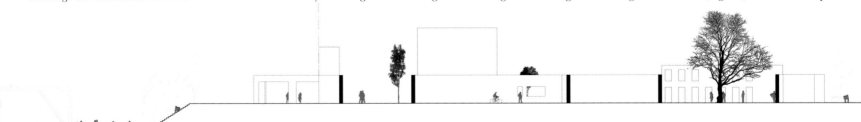

图17 Event 17　剖面显示封闭空间的交叉 Section showing the crossing of enclosed worlds

图18 Fig.18　从地铁步行到卧室 Walking from metro to bedroom

图16 Fig.16　鸟瞰图 Birds' eye perspective

参考文献 References

1. 理查德·塞尼特，（2006年），城市时代的"开放城市"，德国柏林，HYPERLINK "http://www.urban-age.net" www.urban-age.netRichard Sennett, 'The Open City', www.urban-age.net, 2006 (accessed July 2012)
2. M. Christine Boyer，（1994年），《具有集体记忆的城市：历史图像和建筑娱乐》，美国马萨诸塞州麻省理工大学出版社 M. Christine Boyer, The City of Collective Memory: Its Historical Imagery and Architectural Entertainments, Cambridge, MIT Press, 1994
3. Guy Oliver Faure，（2008年），中国社会及其新兴文化，《当代中国期刊》，第17卷，第469-491页 Guy Oliver Faure, 'Chinese Society and its New Emerging Culture', Journal of Contemporary China, vol. 17, no. 56, August 2008, pp.469-491 and George Chu Sheng Lin, 'Chinese Urbanism in Question: State, Society, and the Reproduction of Urban Spaces', Urban Geography, vol. 28, no. 1, 2007, pp.7-29
4. 陈飞和Ombretta Romice，（2009年），通过类型学方法保护城市设计中中国城市的文化认同，《国际城市设计》，第14卷，第36-54页 Chen Fei and Ombretta Romice, 'Preserving the Cultural Identity of Chinese Cities in Urban Design Through a Typomorphological Approach', Urban Design International, vol. 14, no.1, 2009, pp.36-54
5. 缪朴，（2011年），美好新城：自20世纪80年代起中国城市公共空间存在的三大问题，《城市设计期刊》，第16卷，第179-207页 Pu Miao, 'Brave New City: Three Problems in Chinese Urban Public Space Since the 1980s', Journal of Urban Design, vol. 16, no. 2, 2011, pp.179-207
6. Jung Byung-Eon，（2011年），"贾樟柯的电影《小武》中空间资本化及其无地方性"，문학과영상 [韩语]，2011年第12卷第2期，第4页 Jung Byung-Eon, The Capitalization of Space and its Placelessness in Jia Zhangke's Pickpocket, The World, and Still Life, 문학과영상 [Literature and Video] [in Korean], Vol. 12, No. 2, 2011, p. 4
7. 朱小地（2000年），Die Entwicklung und Erforschung des öffentlichen Raums in China. In Koegel, E., editor, "中华人民共和国和德国就公共区域的对话。2000年3月在深圳召开的建筑学和城市发展第二届德中研讨会文件"，第117-122页Zhu Xiaodi, 'Die Entwicklung und Erforschung des öffentlichen Raums in China', in Eduard Koegel (ed.), Dialogues About Public Areas in the PR China and Germany. Documentation of the 2nd German-Chinese Symposium on Architecture and Urban Development in Shenzhen in March 2000, 2000, p.118Maurice Merleau-Ponty, The World of Perception, Cambridge, Cambridge University Press, 2004 Li Xiaodong and Yeo Kang Shua, Chinese Conception of Space, Singapore, China Architecture and Building Press, 2007
8. 梅洛·庞蒂，M.（2004年），《知觉世界》，剑桥大学出版社Maurice Merleau-Ponty, The World of Perception, Cambridge, Cambridge University Press, 2004
9. 李晓东和杨茳善，（2007年），中国空间知觉，新加坡，中国建筑工业出版社 Li Xiaodong and Yeo Kang Shua, Chinese Conception of Space, Singapore, China Architecture and Building Press, 2007
10. 缪朴，（1990年），中国东南部传统城市的七大特征，《传统住宅和居住点概览》，1.2 Pu Miao, 'Seven Characteristics of Traditional Urban Form in Southeast China', Traditional Dwellings and Settlements Review, 1.2, 1990
11. 李晓东和杨茳善，（2007年），《空间的中国观念》，新加坡Li and Yeo, Chinese Conception of Space, Singapore, 2007
12. 缪朴，（1990年），中国东南部传统城市的七大特征，《传统住宅和居住点概览》，1.2，第39页 Pu Miao, 'Seven Characteristics of Traditional Urban Form in Southeast China', Traditional Dwellings and Settlements Review, vol. 1 no. 2, 1990, p.39
13. 李晓东和杨茳善，（2007年），《空间的中国观念》，新加坡Li and Yeo, Chinese Conception of Space, Singapore, 2007

Chinese Characteristics'. It is a framework that is formed by the existing characteristics of the location, and holds an interpretation of fundamental perceptual principles of Chinese culture. It offers a series of enclosed worlds with humanly scaled courtyards, gardens, and small open areas. Walls are the most prominent physical manifestation. The final design looks, from a birds-eye perspective, rather chaotic[Fig. 15 and Fig. 16], but eye-level perspectives show a movement through a clearly understandable series of enclosed worlds[Fig.17 and Fig.18].

Contemporary Interpretation

The urban design project presented in this paper aims at a contemporary interpretation of some key principles of the Chinese urban tradition. The project shows that they can again form the durable elements of the city, which can continuously absorb new functions and patterns. Enclosure, with the wall as its symbol, forms the key element in this, thus mysteriously presenting space little by little. It touches the core of the Chinese perception of space by creating and ordering empty volumes. Being enclosed by architectonic elements, space can become a particular 'place' that is meaningful for everyday life, offering individual and collective local identity.

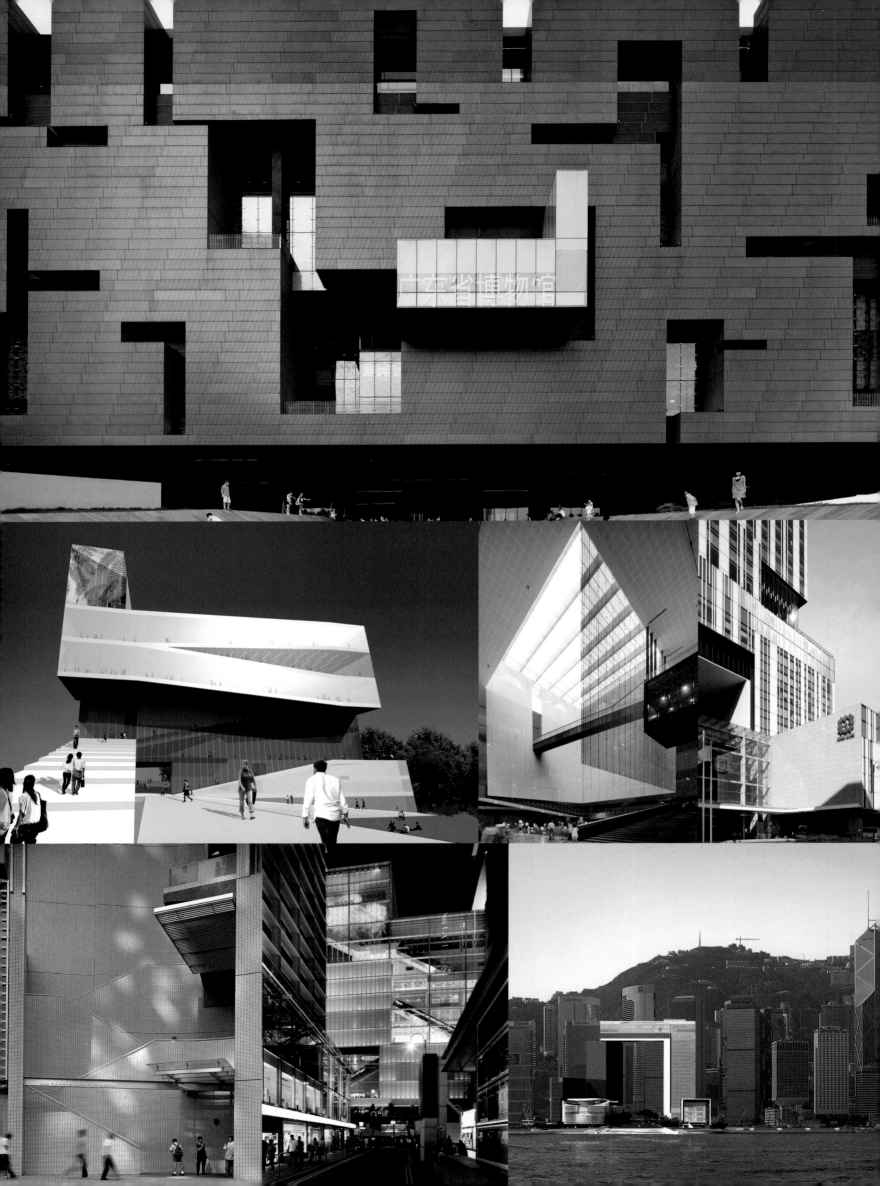

主题 IV：
案例分析 3/4
**Subject IV :
project discussions**

严迅奇
许李严建筑师事务有限公司执行董事

Rocco YIM
Executive Director
Rocco Design Architects Limited
Hong Kong, China

避免奢侈的分离
Avoiding the Wasteful Disconnect

是建筑塑造一座城市，还是城市影响建筑呢？这个问题其实并不好回答。发展中的城市趋向于期望依靠建筑物来提升城市形象。然而，各处又都有其根深蒂固的内在传统、文化、生活方式以及自然属性等要素。如果建筑可以重新发现并审视这些抽象的、看似难以捉摸的要素，并且激发这些要素以创造出与文化和城市化相关的实体，它们就可以牢牢把握并引导眼下城市形态的发展，从而塑造未来的城市面貌。相反，在高度发达的城市中，例如香港，当代的城市力量倾向于强化甚至超越内在文化特性的影响。基础设施配置、行为模式、运动以及空间和结构体系会产生有形无形的张力，影响建筑的发展。此种情况下，是城市在影响着建筑。如今，对于亚洲高密度的大都市来说，最关乎其发展的问题无疑是如何确保城市与建筑之间的创造性互动，而不是费时费力地将城市与建筑割裂开来。

Does architecture shape the city, or does the city shape architecture? This is a question that defies simple answers. Developing cities tend to look toward architecture to promote their identity. However, ingrained in every place, there are inherent elements of tradition, culture, way of life, and natural attributes. If architecture could rediscover these abstract, seemingly elusive elements, evoking them to create culturally and urbanistically relevant entities, they could anchor and induce the development of the immediate urban form and help shape the future city. Conversely, in a highly developed city such as Hong Kong, contemporary urban forces tend to reinforce and sometimes transcend the influence of inherent cultural qualities. The infrastructural configuration, the behavioural patterns, the movements, and spatial and structural systems exert both invisible and tangible tensions and contribute to the evolution of the architecture. The city, in this instance, helps shape the architecture. How to ensure creative interaction rather than wasteful disconnect between the city and architecture is surely the most pertinent question facing Asia's high-density metropoles of today.

上海九间堂

我们应邀在上海郊区浦东设计一组低密度住房。上海的这个地区非常洁净、整齐、具有田园风情——是一座典型的中国城市，但却完全不令人兴奋。我们找不到这座城市能"给予"我们任何东西。因此，我们追溯到了它的历史。我们考虑了本地的庭院住宅并设计了一个类型。虽然该类型来源于历史但却适合于当代中国家庭。他们有子女——一个或两个。他们有需要赡养的父母，也会有朋友。只要他们愿意，他们想大家住在一起。因此这种像过去那样在墙内修建住宅的想法是了一个可行的类型。

所以我们从围墙、轴线、开放空地和沿着轴线的功能空地入手。最后得到的是一种具有适应现代中国家庭的空间结构的含蓄型住宅[图1-图2]。从入口庭院可以看到朝向后花园的居住区和远处的主要住处。双容量的生活区向后朝着入口。卧室和独立花园单独留在一边从而使得每个家庭成员都有尽量有自主性，同时在需要时能进行公共生活。这是非常适合中国现代富裕家庭的住宅。但是就城市规化而言，它并未从城市中吸取任何东西，也对城市没有任何反馈。

广东博物馆

广州的新珠江新城区包含了住宅、商业和公共建筑。主轴南北贯穿。在南端的是绿化轴。在那里，你能找到由剧院、儿童中心、图书馆和博物馆组成的文化区。当我们参与博物馆设计大赛时，我们考虑到了几点。其一，当然是它应当与这一建筑群相互作用从而形成一个区域。扎哈·哈迪斯的歌剧院已经设计出来，所以我们有除了网格之外的一定的环境。图书馆还没被设计出来。我们以百宝箱（来源于中国传统文化遗产）的比喻开始。百宝箱是装贵重物品的容器。而箱子本身作为容器同样是有价值的。但是城市理念认为箱子之所以能穿过绿化带与歌剧保持着交流，同时也吸引着建筑外面的人们，是因为整个箱子漂浮在一个波状的下底之上[图3]。我们在那次大赛中胜出，随后受邀参加另外一个设计图书馆的比赛。我们认为图书馆是同时集正形空间和负形空间于一体的，具有连续的跨越各个层面的空间运动。但更为重要的是，图书馆必须形成文化接头的一部分，通过可见的活动为歌剧院、博物馆和图书馆之间的空间注入活力；负形空间或半私人的开放空间与公共空间相融合并构造出相邻的重要建筑的景色。所以我们所希望的是，图书馆、博物馆以及穿过马路的歌剧院和儿童中心能够为城市环境设定一个基调。它将具有一种文化氛围、充满生气、活动丰富并与建筑风格浑然一体。

博物馆于2011年年底竣工。基本上达到了建筑的设计意图——这是一个散发着神秘与激情的实体，很像一个百宝箱，吸引着人们进入并探索里面有什么希奇。一旦进入了，你能看穿这些空间与循环元素的不同层次[图4]。展览在非正式的循环空间、大体积以及穿过建筑正面的开口处进行，填补了画廊之间的空隙——例如，通过远处的歌剧院的风景。根据预期的设计效果，半公共空间渗入到博物馆范围内。希望开放式的通道在未来能够保持运转。

但是就本区域的规模而言，建筑物为城市空间带来活力并为该区域如何发展提供提示的基本意图似乎并未实现。其他人设计了中心轴，打破了二者之间的视觉和空间渗透性和流动性。我们没有在图书馆设计大赛中胜出；该图书馆由另外一个建筑师设计。它具有一个非常构造性的、很不透明的外观。这些建筑变得就像置于贫瘠的空间中的物体一样。这并不是建筑与城市之间理想的关系。而不幸的是，这样的情况正在整个中国上演着。

荷李活华庭

香港非常与众不同。它具有充满生机的公共领域，包括街道、台阶、桥梁以及自动扶梯系统等。早在十多年前，我们在这周围一带设计了一个很小的项目——一些高密度的住房[图5]。它位于上环的老区。城市这部分的容积率平均在9到15，不包括绿地容积。所以我们必须要修建高楼大厦。这两座高层建筑都非常简单；它们完全一样但是具有实用性的高楼满足了准确定位的基本功能，而且没有忽视空气对流等问题。

该设计的挑战在于它们如何接触地面。环绕该地区的两条路的水平距离为20米。此外，这两条路承受了大量的交通量。所以我们利用这个机会在一系列公共和私人空间里制造出一系列的公共和私人路径从而允许城市通过发展进行渗透。一条公共线路穿越了该地区，并上升至后面的住宅区。一条高层建筑的私人走廊为居民提供了通道。通过不同层面，公共开放空间被私人开放空间点缀并缠绕其间。安防和安全在视觉上看存在一个基本的问题（因为这是住宅开发），但实际上不存在。公共和私人的桥梁在视觉上相接但实际上并没有相接。

所以有人从公共桥梁上经过，也有人从高层建筑经过。这整个住宅开发被包括在五彩斑斓的居住区之中。这是一种非常可持续的发展，因为它实际上让大多数人能够从中层走到中央区的工作场所而不用乘坐公共汽车或出租车。这变成了一种适于步行的发展。这就是我们在城市中如何利用密度以及建筑如何反映城市的需要。

国际广场

国际广场占领了一个非常与众不同的环境，是一个具有不同性质的项目[图6]。这是在尖沙咀最繁忙的部分进行的商业开发。该地区以前有一座酒店，但是尖沙咀这个地区的地价已经很高了，所以如果再在这里修建另一座酒店从经济上来说没有什么意义。该地区打算发展成为一个集商店、餐厅和影院于一体的零售目的地。因此我们必须设计出一套将所有东西以某种方式堆叠在一起从而能沟通不同层次的战略并带来愉悦的经历。

在这个过程中，我们在建筑原地的一侧修建了一个迷你广场。我们把香港地铁和地下通道直接连接了起来。我们建造了一个具有堆叠层面的建筑，提高了餐饮服务楼面以便他们能看到视野极佳的穿越海港的天际线。零售店，影院和餐厅可以通过一系列快车和当地的自动扶梯以及分散的中庭空间到达，这些中庭也因此成为了穿透综合建筑不同层面的垂直客厅。

来自这座城市不同地区的人们可以领会到这座建筑是如何运转以及如何发展的。在其内部，你能感受到这座城市的脉搏和能量——它们似乎已经渗透到了这座建筑中。在其高层，你能穿过海港与香港岛建立起联系。这座建筑的能量来自这座城市并辐射入这座城市——以一种互利的方式。

理大教学酒店综合大楼

这座香港理工大学的教学酒店综合大楼是另外一个例子[图7]。该地区一侧为繁忙的交通隧道，另一侧为娱乐区。我们需要在这一有限的区域放入以下三部分：酒店、学校以及教职员宿舍。混合使用和功能叠加在香港是很常见的。在这样的项目中，你必须确保不同的功能之间不会有冲突并且一些可兼容的功能能够以积极的方式相互作用。

因此，我们决定将教职员宿舍放到公共开放空间旁边——面向相对安静的环境。我们将教学设施放到较低层和地下室中。至于酒店，我们把房间放到顶层并将公共功能安排到了底层，如此一来，后者可以由学校和酒店共享。我们同样确保了本区域两侧的视觉渗透性和透明度，并且我们再次提供了视野——在旧的高程基准之上——从而所有酒店房间，以及健身俱乐部、水疗和公共会议区都能获得周围环境以及香港岛的连续的视野。本项目处理了功能要求和城市脉络，并且它给予了这样一个密集空间所需的真实的和视觉的渗透性。本建筑的塑造主要由这座城市、方向、入口应当如何安置、高度、以及出于酒店哪个位置具有最佳视野的考虑所决定。但是同时，它建立了一种形式，让将来在其周围的开发能够与其相联系。这将带来一种形式上的一致性和地域感。

香港理工大学设计学院

香港理工大学的校园严格意义上说来不在这座城市中，但是校园本身却像一座城市。在香港，即使是校园中的某个地点都是被约束的。这个校园受到现有的约束有：运动设备、马路对面必要要与其连在一起的另一个大学的开发，以及这栋建筑的公共领域与校园本身的公共领域之间的连接。

我们对设计学院大楼比赛提出的想法是设计学院的不同功能（图书馆、画室、工作室和美术馆），私人和更多的公共区域都尽可能不要水平分层，而应该为一个连续的空间实体[图8]。我们建造了一个公共环路，它穿过校园中央大厅，围绕足球场，（通过一系列坡道）穿过礼堂、展览空间以及工作室。学生上楼进入按照连续空间环路排列的不同的工作室中。从工作室出来后，学生会经过公共道路下方进入校园的另一部分。

这是一种交互系统。建筑与其周边环境交流，能促使设计学院更加民主、开放和易于理解。

添马舰开发项目

添马舰是香港特别行政区政府总部。它容纳了政府的十四个部门，因此在某种程度上一定要有稍微的装饰。香港政府所有驻地都位于这一个位置。其独特之处在于，将协调三个非常不同的政府机构到一个位置[图9]。在它们之间进行了虚拟对话后，我们借此机会带来了一个贯穿中央的连续的公共空间，将城市的中心部分与未来海滨连在了一起。这是一个自西向东的连续长廊。公众能24小时不间断地通过中心。进入政府大楼的入口在侧面，因此连接

图3 Fig.3　广州广东博物馆外立面
Guangdong Museum, Guangzhou – exterior façade (photo by Almond Chu)

图1 Fig.1　上海九间堂低密度住宅庭院
Jiu Jian Tang low-density housing, Shanghai – courtyard

图2 Fig.2　上海九间堂低密度住宅街道立面
Jiu Jian Tang low-density housing, Shanghai – street façade

部分和未来海滨真正成为了该建筑的主要驱动力。我们希望这个重要的、（目前）唯一的将这座城市与海滨南北相连的连接能成为连接香港目前杂乱的开放空间与公园的催化剂。我们可以建立相互连接的绿地公园网[图10]，用添马舰树立第一个榜样。我们希望我们在建筑上的努力能鼓励更加有意义的城市姿态出现。

西九龙文娱艺术区

最后介绍一个地区，它并非一个建筑。我们参加了西九龙文娱艺术区总体规划的最后阶段设计[图11]。该地区将容纳连同配套设施在内的十四个文化场馆。但是在香港，这样的区域像独立的建筑一样，已经被这座城市广泛熟知。

这个位置受不同情况的约束。穿过海港，你可以看到香港的天际线和它不同的地标。朝北望去，你能看到九龙站所有不同的建筑区。在其对面，有一片住宅区和酒店。在其东侧是处于高地上的九龙公园。朝着东北部九龙老区望去，我们能看到市场与小贩的活动等等。

因此，我们创造出了一个在各个方向上都具有意义深远的视觉联系的地区，甚至穿越海港。我们建造了三个区域。面对海港的绿地坪是一条波浪形的路径和一系列覆盖在商店与餐厅的绿色房顶。它朝着东方蜿蜒而行最终与九龙公园相连。该地区的一些区域互相突出已形成一个相互协调的整体。九龙现有的网格也被带入了该地区、创造出纹理与可走性——这在香港和九龙都非常典型。

每条这样的街道都有旋转轴以便行人可以与对岸海港的某个特别的地标建立视觉联系。

将一切统一在一起的是不同的公共场所——广场、露台、街道。环境受到城市脉络的推动。在这些环境内，我们在创造可能为未来建筑物带来活力的条件。

奢侈的分离

因此问题在于：我们如何避免奢侈的分离并确保我们在建筑与城市之间具有创造性的互动？这是我们在这个时代所面临的最切中要害的问题。

图4 Fig.4　广州广东博物馆室内
Guangdong Museum, Guangzhou – interior (photo by Marcel Lam)

图5 Fig.5　香港荷李活华庭高密度住宅
Hollywood Terrace high-density housing, Hong Kong

图6 Fig.6　香港尖沙咀iSQUARE商业开发
iSQUARE commercial development in Tsim Sha Tsui, Hong Kong (photo by Marcel Lam)

图7 Fig.7　香港理工大学教学酒店综合大楼
Teaching Hotel Complex for Hong Kong Polytechnic University (photo by Marcel Lam)

Jiu Jian Tang, Shanghai

We were asked to design a set of low-density houses in Pudong, in the suburbs of Shanghai. This part of Shanghai is very clean, orderly, idyllic – a typical Chinese city, but totally uninspiring. There was nothing we could find that the city could 'give' us. So we turned back to history. We looked at vernacular courtyard housing and devised a typology which, although originating from the past, is suitable for the contemporary Chinese family. They have children – one, maybe two. They have parents who they have to support, and they may have friends. They would want everybody to live together for as long as they are willing to. So the idea of a dwelling within a wall, like in the old days, is an applicable typology.

So we started with enclosures, axis, open space, and functional space along the axis. We ended up with an introverted house [Fig. 1 and Fig. 2], with a spatial organisation that is tailored to the modern Chinese family. The entry court looks through the living area toward the back garden and the master quarters beyond. The double-volume living area looks back towards the entrance. The bedrooms and independent gardens are banked along one side so different family members can have their own autonomy as far as possible, as well as communal living when the occasion demands. It is a very suitable accommodation for the modern wealthy Chinese family. However, it takes nothing from the city, and gives nothing back to the city as far as urbanism is concerned.

Guangdong Museum

The new Zhujiang Xincheng district of Guangzhou comprises residential, commercial, and public buildings. The main axis runs from the north to the south. At the southern tip is a green axis. There, one finds the cultural quarter, comprised of the opera, the children's centre, the library, and the museum. When we entered the design competition for the museum, we had several things in mind. One, of course, was that it should interact with this group of buildings to form one district. Zaha Hadid's opera house was already designed, so we had some context apart from the grid. The library was not yet designed.

We started with the metaphor of a treasure box (originating from traditional Chinese heritage), the box being the container of valuable objects. The box itself is always as valuable as the container. But the urban idea is that this box maintains a dialogue with the opera across the green belt, while also drawing people in beyond the exterior of the building, because the whole box floats above an undulating lower base [Fig. 3].

We won that competition, and were then invited to take part in another competition for the library. We perceived the library as an integrated mass of both positive and negative spaces, with sequential spatial movement across levels. But more importantly, the library was meant to form a part of that cultural piecing, energising the space between the opera, museum, and library, with visible activities; both the negative, or the semi-private open space fusing with the public space, and framing views of important buildings in the neighbourhood. So the hope was that the library, museum, and opera across the road – along with the children's centre – would set the tone for this new urban setting. It would have a cultural ambience. It would have something vibrant, something activity-based, and integrated with the architecture.

The museum was completed at the end of 2011. The design intent in the architecture was basically achieved – it is an object that exudes mystique and intrigue, rather like a treasure box, to attract people to go inside and explore what it is holding. Once inside, one can see through these different layers of space and circulation elements [Fig. 4]. Exhibitions are held in the informal circulation zones, large volumes, and the openings all across the façade, providing relief between galleries – for instance, with a view of the opera in the distance. As intended, semi-public space seeps into the confines of the museum. Hopefully the open passageway will stay in operation in the future.

图8 Fig.8　香港理工大学设计 学院竞赛方案
Competition proposal for the School of Design, Hong Kong Polytechnic University

图9 Fig.9　香港添马舰特区政府总部
Tamar Development, the government headquarters of the Hong Kong SAR

But at the scale of the district, the general intention of the architecture energising the urban space and providing hints of how the area could develop, doesn't seem to come out. Somebody else designed the central axis, and the visual and spatial permeability and fluidity between the two was broken. We didn't win the competition for the library; it was designed by another architect. It has a very tectonic, but very opaque façade. The buildings have become like objects sitting in a sterile space. This is not an ideal relationship between architecture and the city. And this, unfortunately, is happening all over China.

Hollywood Terrace

Hong Kong is very different. It has a vibrant public realm of streets, steps, bridges, and escalator systems. In this neighbourhood, more than ten years ago, we designed a very humble project – some high-density housing [Fig. 5]. It is in the old district of Shang Wang. The plot ratio in this part of the city averages about 9 to 15, excluding the green plots. So we really had to go high rise. The two residential towers are fairly simple; they are identical but practical towers satisfying the basic function of correct orientation without overlooking, with cross-ventilation and so on.

The design challenge was how they met the ground. Two roads surrounding the site have a level difference of 20 meters. What's more, the roads carry a substantial volume of traffic. So we made use of the opportunity to create a series of public and private routes in a series of public and private spaces that allows the city to permeate through the development. A public route crosses the site, and ascends to the residential neighbourhood at the back. A private lobby for the residential tower provides residents with access. Across the different levels, public open space is interspersed and intertwined with private open space. Visually, not physically (because it is a residential development), there is this basic concern for security and safety. Public and private bridges meet visually but don't interact physically.

So people come from the public bridges and people come from the residential towers. The whole development is subsumed into a very colourful neighbourhood. It is a very sustainable development in that it actually enables most people to walk to their workplaces from the mid-levels to the central district, without taking the bus or a taxi. It becomes a walkable development. That's how we can make use of density and how architecture can inform a need in the city.

iSQUARE

iSQUARE occupies a very different context, and is a project of a very different nature [Fig. 6]. This is a commercial development in the busiest part of Tsim Sha Tsui. A hotel formerly stood on the site, but land prices in this part of Tsim Sha Tsui have become so high that to build another hotel was probably not going to make any sense economically. It was to be developed as a retail destination with shops, restaurants, and cinemas. So we had to devise a strategy of stacking things up in a way that would allow access to the different levels and encourage an enjoyable experience.

In the process, we set the building back to create a mini plaza on one side. We brought in a connection directly from the MTR to the basement. We created a building of stacked datums and elevated the F&B floors so they have excellent views of the skyline across the harbour. The retail, cinema, and restaurants are accessed by a series of express and local escalators and distribution of atrium space – i.e. vertical living rooms that penetrate the different levels of the complex.

From different parts of the city, people can appreciate how the building works and see how it moves. From the inside, you feel the pulse and energy of the city, which seems to permeate into the building. And at the upper level you have a relationship with Hong Kong Island across the harbour. The energy of the building comes from the city and it also radiates into the city, in a way that one actually benefits the other.

Teaching Hotel Complex

This teaching hotel for the Hong Kong Polytechnic University is another example [Fig. 7]. The site has a busy traffic tunnel on one side and an entertainment district on the other side. We needed to put three components on this confined site: a hotel, a school, and staff quarters. Mixed-use and the stacking of functions is not uncommon in Hong Kong. In such projects, one must make sure that the different functions are not in conflict, and that some of the compatible functions can interact with each other in a positive manner.

So we decided to put the staff quarters next to a public open space – facing a relatively serene environment. We put the teaching facilities in the lower level and in the basement. As for the hotel, we put the rooms on top and the public functions at the base, so that the latter could be shared between the school and the hotel. We also ensured that we had visual permeability and transparency between the two sides of the site, and and that we made the view available again – above the old height datum – so that all hotel rooms, as well as the health club, spa, and public conference area could have an uninterrupted view of the surroundings and of Hong Kong Island.

The project addresses the functional requirements as well as the context of the city, and it allows the actual and visual permeability that such a dense space needs. This building is shaped, principally, by the city, by the orientation, by where access should be placed, by the heights, and by consideration of where the view is best for the hotel. But at the same time, it sets a form that future developments around it will be able to relate to. This should lead to a form of identity and sense of place.

School of Design, Hong Kong Polytechnic University

The campus of the Hong Kong Polytechnic University is not strictly speaking in the city, but the campus is like a city. And in Hong Kong, even a site within a campus is constrained. This one is constrained by existing sports facilities, by the presence of another university development across the road that must be linked, and by the linkage of the public realm of this building with the public realm of the campus itself.

The idea we proposed in the competition for the School of Design building is that the different functions in the design school (the library, studio, workshop, and art gallery) – both the private and the more public areas – shouldn't be stratified horizontally, as far as possible. They should really be one continuous spatial entity [Fig. 8]. We created a public loop that travels in from the campus concourse, around the football pitch, and (via a series of ramps) through the auditorium, the exhibition spaces, and the workshops. The students ascend to access the different studios, which are arranged in a continuous spatial loop. From the studios, students pass beneath the public road to go into the other part of the campus.

It's an interactive system. The building speaks with

图10 Fig.10　香港添马舰特区政府总部
Tamar Development, the government headquarters of the Hong Kong SAR
(photo by Lam Pok Yin, Jeffrey)

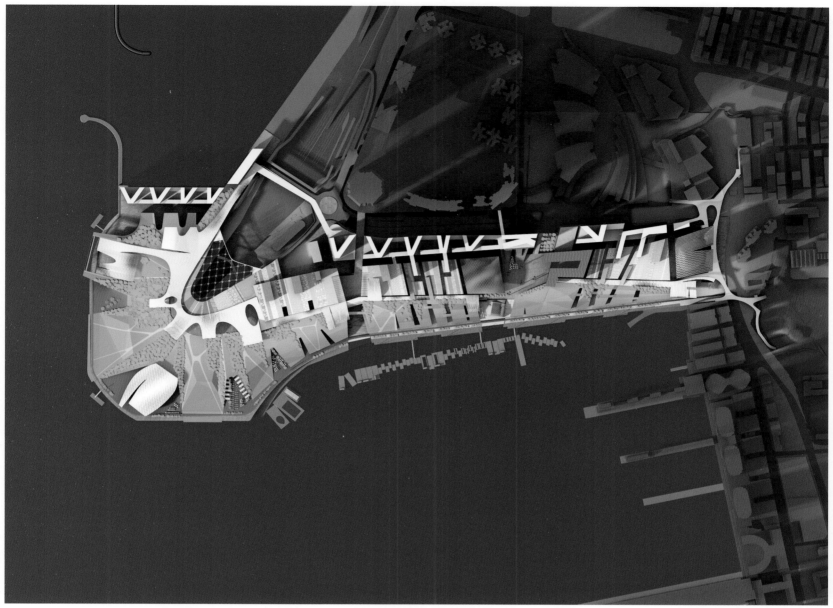

Fig.11 West Kowloon Cultural District master plan

the surroundings. It encourages a design school that is more democratic, open, and understood.

Tamar Development

Tamar is the government headquarters of the Hong Kong SAR. It houses the fourteen bureaus of the government, so had to be – in a way – slightly ornamented. All the seats of government of Hong Kong are located at this one site [Fig. 9]. The unique feature is the accommodation of three very different bodies of government on one site. With a virtual dialogue between them, we took the opportunity to bring an uninterrupted public space right through the centre, linking the inner part of the city to the future waterfront. This is a continuous promenade from the west to the east. The public has uninterrupted 24-hour access across the centre. The access to the governmental buildings is from the side, so the link and the future waterfront is really the main driving force for the architecture.

We are hoping that this important, and (at the moment) only north-south link from the city to the waterfront is a catalyst for the linking of Hong Kong's currently disjointed open spaces and parks [Fig. 10]. We could have a connected network of green parks, with Tamar setting the first example. We hope that our architectural endeavour will encourage urban gestures of a more meaningful nature.

West Kowloon Cultural District

Finally, a district – not a piece of architecture. We took part in the final stage of the design of the master plan for the West Kowloon Cultural District [Fig. 11]. The district is to house 14 cultural venues, along with supporting facilities. But a district, like individual pieces of architecture in Hong Kong, is very much informed by the city.

The site is bound by varying conditions. Across the harbour, you see the skyline of Hong Kong and its different landmarks. Looking towards the north, you see all the different developments on top of Kowloon Station. Opposite, there is a residential area and hotel. On the eastern side is Kowloon Park, which is one level above ground. And looking towards the old part of Kowloon in the north-east, we see the activities of the street markets, hawkers, and so forth.

So we created a district that has meaningful visual connections on all sides, and even across the harbour. We created three zones. The green terrace, facing the harbour, is an undulating path and a series of green roofs over shops and restaurants. It travels eastward and will eventually link up with Kowloon Park. Some of the district's zones protrude into each other to form an integrated whole. The existing grid of Kowloon is also brought into the site the create texture and walkability, which is so typical of Hong Kong and Kowloon. Each of these streets has the axis rotated so that the pedestrian can form a visual connection with one particular landmark across the harbour.

Unifying everything is a series of diverse public places – squares, terraces, streets. The setting is driven by the context of the city. Within the setting, we are creating conditions that will possibly energise the future pieces of architecture.

Wasteful Disconnect

So the question is: how do we avoid wasteful disconnect and ensure that we have creative interaction between architecture and the city? That is the most pertinent question we face in this age.

开拓者的新时代
A New Age for Pioneers

主题 IV：
案例分析 4/4
Subject IV:
project discussions

马赛厄斯·霍洛韦奇 & 马休·霍夫曼
宾西法尼亚大学建筑系讲师
霍洛韦奇—库什纳建筑事务所（HWKN）联合负责人和项目经理/商业发展负责人

Matthias HOLLWICH & Matthew HOFFMAN
Lecturers, Department of Architecture, University of Pennsylvania; Co-principal and Project Manager/Head of Business Development (respectively), HWKN

当今是建筑的历史性时刻。随着婴儿潮出生的一代人正在变老，社会趋于老龄化，我们需要让建筑以社会目的重新获得影响力。老龄化的尊严需要被重新恢复。我们不应该只考虑建筑的功能性，而应该将进步组织与新形势的发展和建筑语言相结合。这是一个建筑能够支持社会改革的时代。通过创建一种对年轻文化的平衡力，并且随着我们年龄变老而逐渐对生命尊重更加的渴望，我们便可以做到这一点。

Today is a historic moment for architecture. With a society growing older and baby boomers ageing, we have the opportunity to put architecture back in the driving seat for social purposes. The dignity of ageing needs to be reinstated. We should not consider architecture in purely functional terms, but with the infusion of progressive new organisations and the development of spectacular new forms and architectural languages. This is the moment when architecture can support a revolution in society. We can do this by creating a counterbalance to youth culture through pioneering attributes for a respectful life as we age.

令人感到惊讶的是，在一个一直在与社会主义做着最艰难的斗争的国家，在过去的850天里，三分之一的美国人退居到最社会化的环境中：养老院。对于这一代人，公共生活是一个陌生的不必要的概念。这一概念是在战后郊区激增的情况下提出的，这些郊区有单独划定的土地，在私有和公有之间有清晰的分界线。为什么在生命最后时刻的唯一选择是居住在一座装饰过的医院中？人们在医院的病房与从未见过面的其他2到4个人一起度过他们最后的日子，现在医用窗帘就是他们自己的空间与其他人的空间的分界线标志。这里有两个问题。第一，养老院从来不是而且永远不可能是家，那么为什么要尝试呢；让我们为这一问题做点事情吧。第二，搬进养老院是可以避免的——但怎样避免呢？

BOOM棕榈泉——原型
两年前，我们就开始在加利福利亚沙漠中棕榈泉附近设计一个新的老年社区[图1、图2和图3]。我们选择的地方是一块未开发的沙地——一块方形的空地。我们还邀请了其他9位建筑师与我们一起将社区设计为他们个人对老龄化、生活方式和建筑方面的看法的具体表现形式。

两种刺激因素确定了项目的起点并指引着我们的工作。第一个是需要从根本上重新想象我们在以后日子中的生活方式。第二个是来源于我们对LGBT（女同性恋、男同性恋、双性恋和变性人）社区生活方式的考虑。我们把LGBT社区的成员定义为"持不同意见的人"。许多人不能跟随社会提供的"典型"生活模式，于是创造了他们自己生活方式的原型。把这两种刺激因素综合在一起，我们发现需求和机遇是完美的结合——重塑一个必须体现老龄人自身的老龄化社区。男同性恋社区不满足于所谓的传统设计和典型的沙漠高尔夫。他们在青春时期就尝试了居住在一起的不同方式，最终决定只生活在一个设计与精致而积极的生活方式同步的社区。设计和开发团队必须树立这样的理念：居民可以选择在同一社区里度过他们更成熟的人生阶段。

我们一直在追寻怎样真实地定义贯穿整个BOOM设计过程的"社区"。不久，我们就发现了任何一个在关注"新老龄化"的新社区都需要关注多代人居住。年长的居民想要居住在年轻一代的附近。因此，BOOM的范围扩大到要吸引各个年龄阶段的居民，包括儿童（因为现在许多年轻人都有了家庭），而不是只将重点放在55岁

以上的年龄阶层中的居民。除此之外，该社区并不排除同性恋者，而是一个多元化的社会，向所有的人开放。在这个方面，BOOM富有创意地提出"就地老龄化"的理念，关注的是一种丰富而充满活力的生活而不是退休或回归后的生活。多代人生活社区的天然活力扩展演变成一种关怀所有人的态度。从历史观点上说，同性恋社区一直站在困难时刻互帮互助的前线，并将继续将这种精神发展下去。因此，安全不仅仅来自先进健康设施的承诺，而且因为居民知道他们正居住在一个充满关怀和活力的社区。

随着时间展开的三个核心价值观
我们必须要创造有意义的生活。社区中不能有任何虚假的事情。人们能够表现自己，实现自己的想法和愿望，将他们自己植入社会。为了帮助他们实现这些，我们设想了一个不断更新自身的自我管理实体，更新是通过一组民主参数和能为居民的需要提供一切的建筑来完成。
第二种核心价值观是高级的形式。我们是通过具有先进便利设施的杰出建筑物和特别的经历来达到这种价值观。当今，许多为所谓的"老年人所建"建筑物看起来已经过时、单调、令人压抑。这就助长了老龄化是消

图1 Fig.1 棕榈泉BOOM – 广场 BOOM Palm Springs – plaza

极的、不合时宜的看法。关注高级形式颠覆了对老龄化的认知，使BOOM令人满意。这种价值观转变了公共认知，使人们对在这里居住的人产生了羡慕——根据这种认知，我们建造了一座对抗年龄歧视的建筑。这也是建筑对人们而言又有了一个真正意义的时刻。从历史观点上说，社区的建筑传达的是公司的身份；在BOOM，我们将真实空间的权力转回给属于它的人们。

第三，BOOM与安全的未来有关。在老龄化过程中，有足够多的事情需要亲自开创。在BOOM，我们消除了一切潜在的不安全性，开始采用非歧视方法，包括年龄和性取向（对于LGBT社区，一旦需要帮助时，许多人们就会回到柜子里去——以避免滥用）。非歧视方法深刻地表现在建筑物和设计中。事实上，在我们的案例中，"养老院"是一个健康中心，是社区中最具独特性的未来主义建筑。它使一个强大而现实的社区成为一体，当需要帮助或生活变得有点困难时，它也不会改变其初衷。互帮互助回到了我们人类帮助他人是一种常识的本源。我们经常发现"你帮助我，我帮助你"这种理念很流行。如果有邻居来拜访你，询问你是否一切安好，该多好啊。

对于这三种核心价值观的执行，需要核对许多的细微差别。当邀请所有年龄阶段的人来建设一个充满活力、没有隔离的社区时，我们将从消除任何年龄的限制开始。无论这个人是40多岁、50多岁、60多岁、70多岁、80多岁还是90多岁，年龄都是相对的。BOOM在乎的是社区和个性，不在乎数字。为了支撑这个观点，我们手上需要一个有着与我们理念相同的社区——这也许是我们工作中最艰难的部分。这就是为什么我们开始积极地建立一个网络社区，然后将其修建成真正的社区的原因。网络社区是基于老龄化的灵感，建立一个关于目标、愿望、灵感、恐惧、心愿和欲愿的生动谈话。老龄化是生活的礼物——我们只需按照正确的方法做事就可以了。

未来BOOM的建筑能够凭借各种各样的设计和规划来支撑礼貌的社交和令人惊叹的建筑。然而，每个设计师对10个微型社区的感觉都是很独特的，BOOM的核心价值观就通过公共生活和个人生活之间的独特梯度表现出来。每个住宅的设计都是公共空间不同程度的复杂结合。每个家的内部和家中每个房间，居民都可以在较大的共享空间直接地来回走动。这创建了一系列从共有到私有的空间，并如他们所愿要求人们在社区中一起生活。

BOOM 棕榈泉的教训——修缮改建

对BOOM 棕榈泉的反应一直都充满了热情以至于我们在考虑一个"BOOM 社区模式"——为其他BOOM的位置勾勒了一副蓝图[图4、图5和图6]。部分项目扩张的方向是对2008年房地产崩溃的一种反应，这次房地产泡沫破裂在美国留下了一部分各种形状和大小的闲置地产。为了适应BOOM的价值观，我们不禁会问："我们怎样在佛罗里达州迈阿密北部的一块空商业区上修建一个活跃的社区？"或者"我们怎样在维加斯的一块空旷的中高层地区修建一个退休村？"

我们正通过将建筑师的注意力转向一个"社区蓝图"并将这个新计划插入目前的城市结构中来彻底抛弃目前存在的传统而陈旧的建筑和都市生活实践。我们正在探索的发生在棕榈泉项目的插入、关系和分散现正将它们的方式应用于各种各样的环境和类型。

我们正在探索新的技术——社会容器——它通过为社区的形成注入刺激因素来检验目前类型中存在的潜能，同时又反过来促使我们每个人重新定义我们自己的未来。我们是社会生物，为了容许我们自己积极地定义我们的生活，我们需要在一个大的（不断变化的）社会网络中，创造并再创造与我们的需要有直接联系的动态环境。

我们的新BOOM原型之一涉及使用一个非常简单的程序移动，以改变与典型的居住发展相关的一切。我们将改变高层住宅楼的上下部分，并将所有的公共项目移动到最高点。这样做的作用是什么呢？它成就了社区，并为集体住宅提供了建筑中最有价值的公用空间。这是一种简单的移动并伴随着长时间的改变和令人惊讶的建筑机会，最后建筑的顶层不仅仅是装饰，而变成了一种社会使命，具有空间上的幽默和创新。

使命

我们相信今天是建筑界具有历史意义的一刻。随着社会老龄化（一件美好的事情——谁不想活的更久）和婴儿潮时代出生的人逐渐老龄化，为了社会目的，我们有机会将建筑带回重要的位置。这不需要纯粹的功能，但需充满先进的新组织和惊人的新形式及建筑语言的发展。这是建筑可以在社会上支撑一场革命的时刻，为数年之后的你和我创造一种值得尊敬的生活，并营造一个与青年化相抗衡的社会文化。

图片由HWKN (Hollwich Kushner) 提供

图2 Fig.2 棕榈泉BOOM – 鸟瞰 BOOM Palm Springs – aerial view

图3 Fig.3 棕榈泉BOOM – 通道区 BOOM Palm Springs – circulation area

图4 Fig.4 西班牙马拉加太阳海岸 BOOM 鸟瞰 BOOM Costa del Sol, Malaga, Spain – aerial view

图5 Fig.5 西班牙马拉加太阳海岸 BOOM BOOM Costa del Sol, Malaga, Spain – view of settlement

It is surprising that in a country that has fought socialism the hardest, the last 850 days for a third of Americans are relegated to the most socialised environment: the nursing home. Communal living is a foreign and unwanted concept for a generation that was raised in the post-war explosion of the suburbs, along with their separate plots of land and clearly demarcated boundaries between private and public. Why is the only choice for the final moment in life a decorated hospital, where people live their last days in bedrooms with two to four other people who they've never met, medical curtains now translucently marking the boundary between their own space and everyone else's?

There are two problems. Firstly, a nursing home will never and can never be home, so why try; let's do something else with it. Secondly, moving into a nursing home can very often be avoided – but how?

BOOM Palm Springs – The Prototype

Two years ago, Hollwich Kushner (HWKN) began prototyping a new age-related community in the California desert near Palm Springs [Fig. 1, Fig. 2, and Fig. 3]. Our site was a virgin plot of sand – a perfectly square tabula rasa. We invited nine other architects to design the community alongside us as physical manifestations of their own beliefs on ageing, lifestyle, and architecture.

Two catalysts defined the beginning of the project and guided our work. The first was the need to radically re-envision the way we live at an older age. The second stemmed from our consideration of lifestyles in the LGBT (lesbian, gay, bisexual, and transgender) community. We came to define members of the LGBT community as 'mavericks'. Many have prototyped their own lives since they could not follow the 'typical' model that society offers. Putting the two catalysts together, we found the perfect match between need and opportunity – reinventing ageing with a community that had to invent itself.

The gay community will not be satisfied with the traditional design of say, the typical desert golf complex. This community has pioneered different ways of living together in their youth, and they are determined to live only in a community whose design is sync with a sophisticated, active lifestyle. The design and development team would have to foster the idea that residents would choose to transition to their more mature years in the same community.

We have been addressing how to actually define 'community' throughout the whole design process of BOOM. We soon discovered that any new community focused on 'new ageing' needs to be multi-generational as well. Older residents want to live near their more youthful counterparts. So, rather than focus on residents exclusively in the 55+ age bracket, BOOM's scope is extended to appeal to residents of all ages, including children (since many younger members now have families). Moreover, the community would not be exclusively gay, but rather a society that was diverse and open to all.

In this way, BOOM imaginatively embraces the philosophy of 'ageing in place', where the focus is on a rich and active life rather than on retirement and withdrawal. The natural liveliness that comes from a multi-generational community extends to an attitude of caring for all. Historically, the gay community has been at the forefront of helping each other in difficult times, and this will continue at BOOM. Thus, security is not just promised by having state-of-the-art wellness facilities, but by residents knowing they are residing in an active community of caring.

Three Core Values Unfold Over Time

We have to allow for the creation of a meaningful lifestyle. There cannot be anything artificial in the community. People need to be able to express themselves, realise their wants and desires, and embed themselves into a social engine. In order for them to do so, we envision a self-governing entity that constantly refreshes itself through a democratic set of parameters, and an architecture that provides everything the inhabitants need.

The second core value is high style. We accomplish this through extraordinary architecture with state-of-the-art amenities and exceptional experiences. Today, most architecture for the so-called 'elderly' looks dated, institutional, and depressing. This is adding to the belief that ageing is something negative and undesirable. Focusing on high style inverts the perception of ageing, making BOOM desirable. This shifts public perception, creating envy of the people living there – and with that we create an architecture that works with us on fighting age discrimination. This is also the moment when architecture has a real meaning again for the people. Historically, corporate architecture has communicated companies' identities; at BOOM we put that power of physical space back where it belongs – the people.

Thirdly, BOOM is about a secure future. There are enough things that personally need to be pioneered during ageing. At BOOM we eliminate all possible insecurities, starting with tactics of non-discrimination. This includes age and sexual orientation (for in the LGBT community, many people go back into the closet once depending on help – to avoid abuse). Tactics of non-discrimination are deeply embedded in the architecture and programming. In fact, the 'nursing home', which in our case is a wellness hub, presents itself as the most exotic and futuristic building of the community. It integrates a strong and realistic community that does not turn its head when help is needed and the days are a bit harder. Helping each other is something that goes back to the origins of our being – as common sense. We often find that the idea of 'I help you when you help me' prevails. How much better it is if a neighbour drops by to see if everything is ok with you.

For these three core values to work, many nuances need to be calibrated. We start with the elimination of any age restriction while inviting all ages to initiate a lively, non-ghettoised community. Whether one considers the 40+, 50+, 60+, 70+, 80+, or 90+ groups, age is relative. BOOM is about the community and personality, and not about a number. To support this notion, we need to have a community at hand that pulls on the same string with us – and that is possibly the hardest part of our job. This is why we have started to actively build an online community that turns into the physical community once the architecture is built. The online community is based on ageing as inspiration and builds a lively conversation about goals, visions, inspirations, fears, wants, and desires. Ageing is the gift of life – we just have to do it the right way.

The future BOOM architecture empowers through a variety of designs and programming supporting respectful socialising, and stunning architecture. While the ten micro-communities are unique to each designer's sensibility, the core values of BOOM are manifested through unique gradients between public and private life. Each housing design is intricately interfaced with varying degrees of public spaces. From the interiors of each home, and each room within the home, residents can move directly to and from the larger communal spaces. This creates a sequence of zones that goes from public to private and invites the community to live with each other as they please.

Learning from BOOM Palm Springs – Adaptive Reuse

The response to BOOM Palm Springs has been so enthusiastic that we are considering a 'BOOM Community Model' – a blueprint for other BOOM locations [Fig. 4, Fig. 5, and Fig.6]. Part of this expanded direction of the project is a reaction to the real estate crash of 2008, which has left a swathe of vacant properties of all shapes and sizes across the USA. Adapting BOOM values makes us ask, "How do we create a viable community in a vacant strip mall north of Miami, Florida," or "How do we create a retirement village in an empty mid-rise in Vegas?"

We are creating a radical departure from the traditional and banal existing praxes of architecture and urbanism, by turning the architect's attention toward a 'community blueprint' and inserting this new plan back into existing urban fabrics. The same programmatic insertions, relationships, and dispersions we are exploring concurrently in Palm Springs, are now making their way into a variety of contexts and typologies. We are inventing new techniques – social condensers – that examine the hidden potential of existing typologies by inserting catalysts for community formation, which in turn enables each of us to redefine our own futures. We are social creatures, and in order to allow ourselves to actively define our own lives we must create and recreate dynamic environments in direct relation to our own needs within a larger (continuously changing) social network.

One of our new prototypes for BOOM involves using a very simple programmatic move to change everything typical residential developments are about. We turn the residential high-rise building upside down and move all public programmes into the skies. What does that do? It empowers the community, and gives the collective of the residences the most valuable space of the building for communal purposes. It is a simple move with long-lasting changes and amazing architectural opportunities, where finally the top of the buildings are not just decoration, but become a social mission and are spatially playful and innovative.

Mission

We believe that today is a historic moment in architecture. With a society growing older (which is a beautiful thing – who does not want to live longer?) and baby boomers ageing, we have the opportunity to put architecture back into the driving seat for social purposes. It need not be purely functional, but can be infused with progressive new organisation, and the development of spectacular new forms and architectural language. It is the moment when architecture can support a revolution in society, creating a counterbalance to youth culture through the pioneering of attributes for a respectful life for you and me in a few years from now. *Images by HWKN.*

图6 Fig.6 西班牙马拉加太阳海岸 BOOM – 广场BOOM Costa del Sol, Malaga, Spain – plaza

跋 Afterword

今天,当我风尘仆仆的从项目地赶回京城,很欣喜的看到桌上"立体城市研究院"的第一本书稿雏形,这是我期待已久的一份作品。我们是一家致力于探索和实践中国城市化发展新策略——立体城市模式的企业。一方面,我们非常务实地在实践层面一步一个脚印的向前;一方面,我们也关注理想层面更激进的探索,这也是我们作为"亚洲垂直城市"国际设计竞赛暨研讨会赞助企业的一个初衷。生命不息,创新不止。

先有竞赛,后有纪录。因此,我首先要感谢,这系列活动的的主办方新加坡国立大学设计与环境学院,感谢亚洲垂直城市竞赛暨研讨会组委会成员,他们是:CHEAH Kok Ming(主席);CHO Im Sik(联合主席);Cardith HUNG Chung Hey(项目协调);GOH Lay Fong(财务);Marcus WONG(外宣);Philip TAY(物流);Dorothy MAN(IT);NG Wai Keen(出版);TAN Teck Kiam(工作营项目);以及我们的合作赞助方世界未来基金会。由于我们三方的共同努力,尤其是主办方在世界范围内的学术影响力,才能有如此高水平的国际化竞赛。

因此,接下来我要感谢除主办方新加坡国立大学外、参与竞赛的其它九所世界知名院校,它们是来自亚洲的香港中文大学、东京大学、同济大学和清华大学;代表欧洲的苏黎世联邦理工大学和代尔夫特科技大学;以及美国的加州大学伯克利分校、密歇根大学和宾夕法尼亚大学。来自世界不同区域不同文化背景的一流学院的积极参与,增强了我们对于大赛乃至培养发掘未来创新中坚力量的信心。

这本书凝结着三个团队的心血。一个是由于冰主编带领的优恀传媒编辑团队,成员包括出品人葉春曦、编辑总监吴博、编辑张雨辰和杨毅、美术编辑逢苹、李渔和黄洁。另一个是由新加坡国立大学设计与环境学院的教职员组成的编辑团队,他们是伍伟坚副教授、CHEAH Kok Ming 副教授)、CHO Im Sik博士,助理教授、Narelle YABUKA 及Cardith HUNG Chung Hey。还有一方是立体之城研究中心。得益于三个高水平团队的无间合作,我们才获得令人欣喜的成果。

最后,我还要感谢以下"亚洲垂直城市"国际设计竞赛暨研讨会编委会成员:

新加坡国立大学设计与环境学院 王才强教授
新加坡国立大学设计与环境学院 王运起副教授
新加坡国立大学设计与环境学院 符育明副教授
新加坡国立大学设计与环境学院 MALONE-LEE Lai Choo博士
新加坡国立大学设计与环境学院及代尔夫特理工大学 尤尔根•罗斯曼教授
香港中文大学建筑学院 何培斌教授
东京大学新领域创新科学研究院 大野秀敏教授
同济大学建筑与城市规划学院 黄一如教授
清华大学建筑学院 朱文一教授
苏黎世瑞士联邦理工大学 基斯•克里斯蒂恩赛教授
代尔夫特理工大学建筑学院 卡琳•拉赫拉斯教授
加州大学伯克利分校环境设计学院 珍妮佛•沃驰教授
密歇根大学陶博曼建筑与城市规划学院 莫妮卡•庞塞德莱昂教授
宾西法尼亚大学设计学院 玛丽莲•乔丹•泰勒教授

郝杰斌 北京万通立体之城投资有限公司 总经理

Returning to Beijing after a hectic business trip to our project city, it was a delight to find on my desk today the draft of this book, the very first one of the City Research Institute Series.

Vantone Citylogic is dedicated to the exploration and implementation of the new Chinese urbanization strategy, i.e., the GREAT City approach. On one hand, we are progressively moving forward on the solid practical ground. On the other hand, our company is also engaged in the creative and idealistic aspects of our pursuit, which is the main reason why we decided to sponsor the Vertical Cities Asia International Design Competition and Symposium. Where there ' s life, there' s creativity.

It is worthwhile to document the contributors of the programme. First and foremost, I would like to thank the organizer of the VCA activities, School of Design and Environment (SDE) at the National University of Singapore (NUS). Vertical Cities Asia would not be possible without the hard work of the following organizing committee members at SDE: CHEAH Kok Ming (Chair), CHO Im Sik (Co-chair), Cardith HUNG Chung Hey (Project Coordinator), GOH Lay Fong (Finance), Marcus WONG (Publicity), Philip TAY (Logistics), Dorothy MAN (IT support), NG Wai Keen (Publication) and TAN Teck Kiam (Workshop Programme). Also, I would like to thank the co-sponsor, World Future Foundation. The VCA activities have been a joint effort of the organizer and the two sponsors. In particular, we acknowledge the global academic standing of NUS to bring together leading institutions from across the world to taketapart in the programme.

Besides our host university, NUS, my appreciation also goes to the other nine top universities participating in the programme. They are: The Chinese University of Hong Kong, University of Tokyo, Tongji University and Tsinghua University for Asia; Delft University of Technology and ETH Zurich representing Europe; and University of California (UC) Berkeley, University of Michigan and University of Pennsylvania from the USA. With these leading institutions of diverse cultural backgrounds, we are confident in not only the success of the VCA competition itself but also realizing our mission to cultivate future creative talents.

The publication of this book is acollaboration across three parties: Youyi Media led by Chief Editor Ms. YU Bing, including Producer Isa Ye, Managing Editor WU Bo, Editors ZHANG Yuchen and YANG Yi as well as Art Editors PANG Ping, LI Yu and HUANG Jie; the NUS School of Design and Environment publication team, including Associate Prof. NG Wai Keen, Associate Prof. CHEAH Kok Ming, Assistant Prof. Dr. CHO Im Sik, Ms. Narelle YABUJA and Ms. Cardith HUNG Chung Hey; and the Vantone Citylogic Research Center. The synergy of the three parties has resulted in this captivating book.

Last but not least, my gratitude goes to the following individuals from the Editorial Committee of the Vertical Cities Asia International Design Competition and Symposium:

Prof. HENG Chye Kiang, School of Design and Environment, National University of Singapore;

Associate Prof. WONG Yunn Chii, School of Design and Environment, National University of Singapore;

Associate Prof. FU Yuming, School of Design and Environment, National University of Singapore;

Dr. MALONE-LEE Lai Choo, School of Design and Environment, National University of Singapore;

Prof. Jürgen ROSEMANN, School of Design and Environment, National University of Singapore and Delft University of Technology;

Prof. HO Puay Peng, School of Architecture, The Chinese University of Hong Kong;

Prof. OHNO Hidetoshi, Graduate School of Frontier Science, University of Tokyo;

Prof. HUANG Yiru, College of Architecture and Urban Planning, Tongji University;

Prof. ZHU Wenyi, School of Architecture, Tsinghua University;

Prof. Kees CHRISTIAANSE, Faculty of Architecture, Swiss Federal Institute of Technology (ETH) Zurich;

Prof. Karin LAGLAS, Faculty of Architecture, Delft University of Technology;

Prof. Jennifer WOLCH, College of Environmental Design, University of California (UC) Berkeley;

Prof. Monica PONCE DE LEON, Taubman College of Architecture and Urban Planning, University of Michigan;

Prof. Marilyn Jordan TAYLOR, School of Design, University of Pennsylvania

Roy Hao
CEO, Beijing Vantone Citylogic Investment Corporation